JN436915

THE ROAD TO THE TOP

SECONDARY BATTERIES

THE ROAD TO THE TOP
SECONDARY BATTERIES

Printed in Seoul, Republic of Korea
ISBN 978-89-521-1846-2 93530

Seoul National University Press
1 Gwanak-ro, Gwanak-gu, Seoul 08826, Republic of Korea

E-mail: snubook@snu.ac.kr
Home-page: http://www.snupress.com
Tel: +82-2-880-5252
Fax: +82-2-889-0785
First Printing: November 30, 2016

THE ROAD TO THE TOP SECONDARY BATTERIES

Jun SONU

TOP: Top of Power

- **T**: Time-to-market
- **O**: "O" (zero) defect
- **P**: Paradigm shift

Preface

The status of secondary batteries has been upgraded from the passive and supporting role of the portable power source to the active role of giving the competitive edge to the portable electronics and electric vehicles. Among various secondary batteries, the lithium ion battery, which is considered as the most advanced secondary battery, has dominated the battery market since the 1990s due to its superior cell performance.

The importance of lithium ion batteries in the Korean battery industry stems from the fact that Korean battery companies entered into the battery business through the development of lithium ion batteries. Up until the early 1990s, Korean companies exceeding minimum standard stipulated by the government in terms of the revenue and number of employees were restricted from entering into the battery business because the battery business was appointed as the business reserved exclusively for small firms. However, Sony's success in the battery business made Korean government as well as Korean companies aware of the potentials associated with advanced batteries like lithium ion batteries. Korean firms which had proven their capability in the area of semiconductor, electronic appliances and chemical business embarked on the development of lithium ion batteries. Needless to say, this significantly galvanized the battery industry in Korea.

The purpose of this book is to give you a clear picture on how the Korean battery industry has been grown to the current position by introducing the development history and the competition dynamics of lithium ion batteries. The lesson we learned from the mistakes we had made in the past will play a pivotal role in reducing the likelihood of making the same mistake repeatedly. I hope this book serves the purpose.

I would like to express my sincere thanks to my old colleagues in LG Chemical and Samsung SDI who had undertaken the establishment of the Korean battery industry for the last two decades. My experience to work together with AEA Technology of the UK and Ford Motors of the USA in 1994-1996 and 2005-2006 respectively made the publication of this book possible.

Finally, my wife and two daughters require a special word of thanks for their support, patience and sacrifice. Their trust on my capability and balanced view enabled me to go on with writing this book.

November, 2016
Jun SONU

Contents

PART 2 LITHIUM ION BATTERIES

PART 3 KOREAN BATTERY INDUSTRY

PART

1

SECONDARY BATTERIES

01
:
Introduction

In Korean people's mind, the 1988 Seoul Olympic Games were the stepping stone to an advanced nation after a lengthy struggle with poverty. After the successful Olympic Games, Korean's expectations on the twenty-first century was very high. Most major firms in Korea prepared the twenty-first century by planning ahead the new business they would pursue to have a sustainable competitiveness. In the midst of the preparation of the future businesses, the news that Sony commercialized lithium ion batteries made a big dent in the Korean firms. In actuality, Koreans were confused at the Sony's entry into the battery business, for the battery business had been defined as the business in which large companies shouldn't enter in Korea simply because the market was too small. Hearing the news of Sony's commercialization of lithium ion batteries, Korean firms realized that the time had come to change their thoughts on the battery business. After serious considerations, they finally included a battery business in the list of the future promising business along with semiconductor, flat panel displays and mobile phones. The business reserved exclusively for small firms was turned into the future core business that would lead Korea to the position of the advanced nation.

Although many companies swamped into the battery business like bees to honey in the 1990s, only a few companies had proven their viability after the elapse of more than two decades. LG Chemical and Samsung SDI not only survived the international competition but also reached the top position in the world battery market.

In retrospect, there were three turning points of risk and opportunity for Korean firms. First, there was inflow of lithium ion battery technology from Japan in 1996. Battery engineers of Sony wanted to grow the battery business as the core business of Sony, but senior managers thought otherwise. There was a conflict in Sony, and senior managers of Sony released battery engineers to suppress conflict. It was Samsung and LG that capitalized on this opportunity. They acknowledged that it was the opportunity they shouldn't miss. Second, financial crisis of Korea in 1998 thwarted all the investment plans on account of the lack of foreign currency. Nonetheless, LG and Samsung decided to go ahead with the construction of the lithium ion battery plants regardless of the financial environment. In contrast, the other Korean companies had given up on the battery business for the difficulties associated with the Korean financial crisis. Third, lithium ion battery makers had undergone field incidents like the outbreak of fire or explosion of mobile phones and laptop computers powered by lithium ion batteries during the period of 2002-2006. They managed to overcome a crisis by a thorough investigation on the cause of the field incidents and subsequent reorganization of their battery business.

With regard to the competition landscape of Korean battery industry, SK chairman's announcement to grow the battery business as the future core business of SK group by participating in the electric vehicle battery business in December, 2007 made not only three-way competition dynamics complete but made future competition more intriguing.

In the early 2000s, Korea was in the bottom rung of a battery business ladder. Korean battery makers moved up to a top rung of a ladder in ten years and now exhibit their strong desire to stay there for a long time.

02

:

Fundamentals of batteries

1) Battery components

The battery is a device that converts chemical energy into electrical energy. For this purpose, four components of the battery, which consists of the anode, cathode, the electrolyte and the separator, are configured to induce the electrochemical reaction. The anode is the electrode in which oxidation reaction occurs during discharge while the cathode is the electrode in which reduction reaction takes place during discharge. The anode acts as the source of electrons, so it is also called the fuel electrode. Metals, which are abundant with free electrons, are used as the anode materials. The anode materials used in batteries encompass zinc, lead, cadmium and lithium. The anode determines the battery characteristics, so the change of the anode material implies the introduction of a new battery system. The cathode is the electrode that accepts electrons and ions during discharge. The oxide and sulfide, which have the capability to hold atoms in the vacant sites of the crystal structure, are used as the cathode materials. The electrolyte is the medium through which ions are transported. Aqueous solution like sulfuric acid or KOH solution is commonly used as the electrolyte, but in the lithium battery, organic solvent is adopted as the electrolyte because lithium metal reacts with water to form hydrogen gas. When the anode and cathode are contacted with each other, heat is generated instead of producing the electricity. Therefore, physical contact between the anode and cathode should be avoided for the successful generation of electricity. The separator located in contact with both the anode and the cathode prevents the electrodes from contacting with each other. The separator plays an additional role of holding electrolytes in addition to the role of separating the anode and cathode. Thickness of the separator is determined by the ionic conductivity of the electrolyte. Ionic conductivity of the organic solvent used in the lithium battery is much lower than that of aqueous solution used in lead acid and nickel cadmium batteries, which makes the separator of the lithium battery much thinner than that of other batteries.

The polymer electrolyte which is used in the lithium polymer battery (LPB) has the dual function of the electrolyte and a separator. Elimination of the liquid component in the battery by the adoption of a polymer electrolyte not only makes the battery safe under the abused condition but also comes in handy in the aerospace application in which the electrolyte leakage should be avoided at all costs.

Battery development activity consists of two parts: the one is to develop a new battery system through the development of new anode materials and the other is to improve the performance of the current battery through the development of cathode materials. When the activities of developing a new battery system outperforms the improvement activities of the current battery, the new battery enters into the picture.

2) Battery characteristics

(1) Voltage and energy capacity

When discussing with battery users like portable electronics designers, their frequent request is to change the voltage. However, change of the voltage is hard to attain unless we change the electrode materials. Based on the battery reaction shown below, we calculate the Gibbs free energy.

$$A \text{ (anode)} + B^{+} \text{ (cathode)} = A^{+} + B$$

The Nernst equation relates the Gibbs free energy with the voltage as follows:

$$\Delta G = -nFE \tag{1}$$

In equation (1), F is the Faraday constant which is equivalent to 96,487C/mole, and n is the number of moles. Once we know the Gibbs free energy of the battery reaction, voltage (E) is determined by the Nernst equation.

One of the most fundamental tests in the battery is the charging and discharging test, which provides us with the comprehensive information pertinent to the battery performance. Discharge curve determines the characteristics of the battery. There are two different kinds of discharge curve, the shape of which depends upon the nature of the battery reaction. In lithium ion batteries, lithium atoms react with the cathode and anode materials uniformly by the process called intercalation, so the boundary between the reactant and the product is obscure. It is called the single phase reaction. In a single phase reaction, the chemical potential changes with the progress of the battery reaction, so the voltage changes with the extent of reaction by the Nernst equation. In contrast to lithium ion batteries, cadmium is transformed into cadmium hydroxide in nickel cadmium batteries, in the process of which a clear boundary between cadmium and cadmium hydroxide is formed. It is typical of two phase reaction, in which chemical potential of each phase should be same, so voltage remains constant by the Nernst equation as long as equilibrium is maintained between two phases.

In summary, the reaction mode makes the difference in the discharge curve. For instance, lithium ion batteries exhibit a slope discharge curve as the voltage changes with the amount of lithium atoms intercalated with the cathode materials. On the other hand, nickel cadmium batteries show a flat discharge curve as the equilibrium is maintained between cadmium and cadmium hydroxide during discharge.

Energy capacity is obtained by multiplying the amount of electric charges with voltage, as follows:

$$\text{Energy capacity (Wh)} = \text{Voltage (V)} \times \text{Electric charge (Ah)}$$

In thermodynamics, voltage is the intensive property whereas the amount of electric charges is the extensive property. Extensive property like the amount of electric charges depends on the size. Thus, energy capacity gives us the information on the amount of electric charges contained in the battery, which determines the running time of the battery. In order to increase the running time, we prefer to choose the battery with high voltage. If the voltage is same, the battery with a large amount of electric charge should be chosen. In the battery, we use energy density and specific energy as criteria associated with the running time. Energy density (Wh/L) and specific energy (Wh/kg) determine how small and how light the battery can be made for the specific application. Size and weight have been an important sales point in portable electronics market. In order to increase the energy density of the battery, we select the anode and cathode materials with high specific capacity (mAh/g) and reduce the amount of inert materials like the current collector and a separator to obtain more space for the active materials. We call the anode and cathode materials active materials to stress that they contain energy to power the portable electronics and electric vehicles.

(2) Rate and temperature performance

The concept of C-rate was introduced to indicate the power capability of the battery. 1C is defined as the current rate at which all the electrical energy in the battery is discharged in an hour. Based on this definition, we can easily see that 2C rate current is ten times higher than C/5 rate current. When the battery is discharged at high C rate, the battery reaches the cut-off voltage prematurely by polarization. We indicate the rate performance of the battery as the amount of the energy to be discharged at a specific rate. For example, if in the battery with 1,000mAh capacity, 700mAh is discharged at 5C rate before it reaches the cut-off voltage, we specify it as 70% at 5C rate. Rate performance is closely related to the impedance of the battery. In order to develop a high power battery, it is imperative to have a full control of the interfacial impedance. Difficulty of the battery development stems from the fact that the interfacial impedance between the battery components cannot be closely controlled.

The reason for the difficulty in getting the engine started in cold winter is because the battery performance declines at low temperatures. The decline in the battery performance is due to the increase in the interfacial impedance at low temperatures. Specifications of temperature performance are similar to those of rate performance. We specify the temperature performance as the amount of energy discharged before the cut-off voltage. The operating temperature range was extended from laptop computers to mobile phones to the electric vehicles. The operating temperature range of laptop computers is 0°C-60°C, which was determined based on the assumption that people usually don't use the laptop computers outside in the cold winter. However, that's not the case in the mobile phones. The mobile phone was invented to use the phone regardless of the location, so people tend to use the mobile phone outside even at low temperatures. Based on this reasoning, operating temperature range of the mobile phones was determined to be -20°C-60°C. However, the charging temperature range of the mobile phones and laptop computers is 0°C-40°C because charging is conducted indoors. Automobiles are used in a very severe condition in terms of the tempera-ture. A car was designed to withstand the hot temperatures of Sahara Desert and the nippy weather of Siberia and Alaska. Therefore, the batteries used in the automobile should have a much wider operating temperature range than the batteries used in portable electronics. The batteries used in the electric vehicle application are required to function properly at the temperature as low as -40°C. The operating temperature range of batteries used in automobiles is -40°C-60°C.

Temperature is arguably the most critical factor to influence the battery performance. As the temperature increases above the ambient temperature, the degradation process is accelerated. Therefore, it is important to take an adequate measure not to expose the battery to the high temperature.

(3) Cycle life

The secondary battery is the battery that can be reused by charging after depletion of energy. It is also called the rechargeable battery. Charging is the process to rejuvenate the battery by injection of electric energy. Energy capacity of the secondary battery, which represents the running time, diminishes with cycles by way of degradation. As the cycle of discharging and charging increases, the energy capacity of the cell decreases gradually.

Degradation of secondary batteries is associated with such phenomena as isolation of active materials, consumption of the electrolyte and gas evolution. There are many ways to define the cycle life depending on the applications, but in general cycle life is defined as the number of cycles at which energy capacity of the battery declines to the level of 80% of the initial capacity. In the hybrid electric vehicle (HEV), the actual current profile at a specific time interval is called the duty cycle, which is equivalent to one cycle in the cycle test. The requirement for the satisfactory cycle life in the HEV is to retain over 60% of the initial power after the elapse of 300,000 cycles.

The concept of depth of discharge (DOD) was introduced to determine the depth of usage of energy capacity. When the total energy in the battery is used in the charge-discharge cycle, we say that the cycle life test is conducted at 100% DOD condition. Needless to say, the cycle life test at low DOD increases the cycle life significantly. For instance, 80% DOD cycle life test doubles the cycle life, and the cycle life increases more than four times when cycle life test is conducted at 60% DOD. In the portable electronics application, the cycle life test is normally conducted at 100% DOD, but in the electric vehicles application, 80% DOD cycle life test is the most commonly accepted test condition.

3) Primary batteries

The primary battery is the one that is discarded when the energy in the battery is used up. That's why it is also called the disposable battery. Figure 1 is the classification of batteries. There is no hard and fast rule to classify the batteries, but it is more convenient to use the anode as the criterion in the primary batteries and to use the electrolytes as the criterion in the secondary batteries because it makes the difference in voltage.

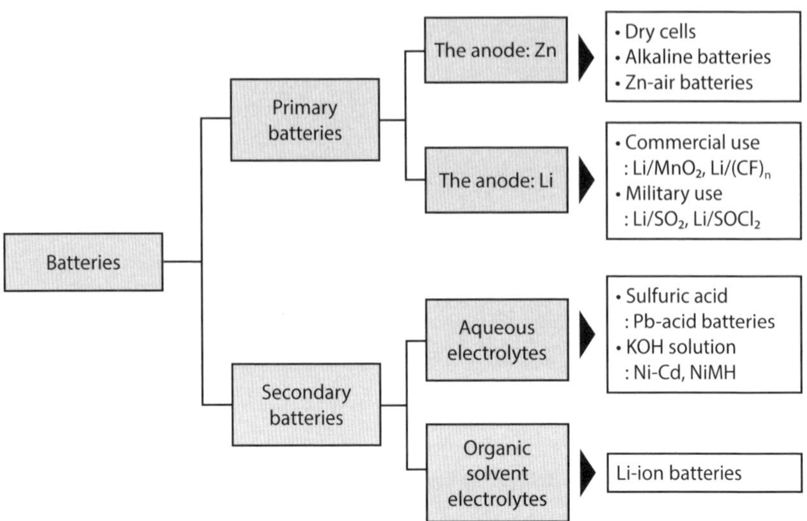

Figure 1. Classification of batteries

(1) Zinc-carbon batteries

The zinc-carbon battery which has been known as the dry cell was invented by French scientist, Georges Leclanché, in 1866. Zinc and manganese dioxide were adopted as the anode and cathode materials respectively, and the carbon rod was used as the current collector of the cathode. The electrolyte, $NH_4Cl_2/ZnCl_2$ aqueous solution, is mixed with carboxy methyl cellulose (CMC) or sodium polyacrylate to make the gell type electrolyte. Voltage of the zinc-carbon battery is 1.5V and it exhibits slope discharge curve. Cup shaped zinc metal is used as the anode, which is fabricated by either deep drawing[1] or impact drawing.[2] The separator is wrapped inside the zinc anode cup, and the inside space of the zinc cup is filled with cathode mixture consisting of manganese dioxide and the

1 Metal sheet is punched on the die to make a cup shaped structure.
2 Coin shaped metal is punched at high speed to make a cup shaped structure.

electrolyte. Finally, the carbon rod is placed in the center of the cathode mixture to form the electron path.

• Alkaline manganese batteries

The effort to improve the cell performance of the zinc-carbon battery gave birth to the alkaline manganese battery, which is just called the alkaline battery in most cases. Alkaline manganese batteries exhibit superior performance in terms of the energy density, rate performance and the shelf life. The major difference between zinc-carbon and alkaline manganese batteries is the electrolyte. Alkaline manganese batteries use high conductivity KOH solution as the electrolyte, so rate capability increases significantly. To meet the high electrical conductivity of the electrolytes of alkaline manganese batteries, zinc sheet is replaced by zinc powders to increase the specific surface area, which results in the change of the internal structure of the battery. Zinc powder mixture is placed in the center of the battery, which we call inside-out structure. The current collector material is also changed from the carbon rod to brass to increase the electron conducting capability. The brass current collector is located in the anode instead of the cathode.

• Mercury-free batteries

Mercury is the essential element in the battery which uses zinc as the anode material. Mercury is added to the anode by spreading Hg_2Cl_2 on the separator. Mercury reacts with zinc anode, forming zinc amalgam layer on the surface of the zinc anode. Zinc amalgam layer inhibits the hydrogen gas evolution by blocking the physical contact of zinc with the electrolyte. Although mercury is the essential element in the zinc anode batteries, it is harmful to the environment. Thus, it was required that mercury be eliminated in the batteries.

In order to find a way to eliminate mercury from the batteries, we need to have the intimate knowledge on what would happen if it were not for mercury. First, a hydrogen gas produced by the reaction of zinc with the electrolyte builds up the internal pressure of the battery, which makes the battery vulnerable to the electrolyte leakage. Second, mercury plays an important role in forming the zinc anode network through the bonding of zinc powder together. Therefore, the zinc anode network would break easily without the addition of mercury when the battery is accidentally dropped on the floor. The breakage of the zinc anode network disrupts the discharge characteristics. Third, it brings about the abrupt drop of the energy capacity in addition to the increase in the self-discharge rate at the later stage of the battery life. Increase in the self-discharge rate results from the soft short, which is caused by the separator penetration of the zinc oxide formed by the corrosion of the zinc anode.

Battery makers were successful in removing the mercury from the batteries by the following method. They added the alloying elements to alleviate the hydrogen evolution. Hydrogen evolution on the zinc surface is related to the surface condition of the zinc, so zinc powders with small surface area were adopted. In addition, new gelling agents were developed to improve shock-resistant characteristics of the zinc. Impurity metals like iron accelerate the corrosion, so manufacturing conditions and environments were to be changed and managed to prevent the accidental inclusion of metallic impurities and to reduce the time to be exposed to the air. Many changes were necessitated to make mercury-free batteries. Development of the mercury free battery is a typical example of the battery development, in which full understanding of the total battery system is required.

As an extension to the alkaline manganese batteries, coin cells were developed to be used in the application of an electronic calculator and a digital watch. Zinc is used as the anode material, and such various materials as mercuric oxide and silver oxide were adopted as the cathode materials in combination with the alkaline solution electrolytes. Unlike alkaline manganese batteries, coin cells exhibit flat discharge curve because of the characteristics of the cathode materials.

• Zinc-air batteries

The adoption of the air electrode in the cathode gave birth to the zinc-air batteries, the main use of which is in the area of a hearing aid. The zinc-air battery has the highest energy capacity among the batteries with the zinc anode. Once the air hole of the zinc-air battery is opened, battery reaction continues without any interruption regardless. The air cathode is made by pressing the active carbon which acts as the catalyst and binder on the metal mesh. The large zinc-air secondary battery belongs to the category of mechanically rechargeable batteries. When the energy inside the battery is used up, zinc anode is replaced by a new one instead of recharging. The zinc-air battery belongs to an air cathode battery family, which encompasses lithium-air and fuel cells along with zinc-air batteries.

There were attempts to apply zinc-air batteries to other applications than the hearing aid by taking advantage of its high energy capacity and mechanically recharging method. In Japan, mechanically recharging zinc-air batteries had been developed for the application in the mobile phone. The zinc anode cartridge was also developed in conjunction with the zinc-air batteries. When we use zinc-air batteries for the mobile phone, we change the zinc anode cartridge purchased from a convenient store or supermarket instead of recharging. Although there was no serious technical barrier relative to the mechanically recharging zinc-air batteries in the mobile phone, the problem was the availability of the zinc cartridge. The interest in the zinc-air batteries for the mobile phone waned with time.

When heavy users of mobile phones such as businessmen and sales persons go to the business trip, the necessity of charging their mobile phones is cumbersome. In order to eliminate this inconvenience, some battery company developed a disposable zinc-air batteries for the mobile phone by taking advantage of its high energy capacity. The disposable zinc-air batteries were in sale in Brazil in 2000-2001. Mobile phone users who need to use mobile phones without the need of charging for a long period of time go to the convenient store and buy a disposable zinc-air battery. After opening the air hole, he or she connects the battery to the mobile phone and use the phone during the business trip.

Daimler-Chrysler had made electric vehicles with a battery company in Israel using zinc-air batteries for the testing and demonstration purpose in 1990s. Zinc-air batteries have a wide range of application from the hearing aid to electric vehicles. In this respect, it is no wonder that zinc-air batteries are considered as one of the primary candidates for the next generation battery.

(2) Lithium batteries

Lithium metal floats on the water because of its low density, 0.53g/cc. The melting point of lithium is 186°C, and it reacts with water and alcohol, both of which have the hydroxide functional group. In order to prevent the inclusion of water during the assembly process, lithium batteries are made in the dry room where the dew point is maintained below -40°C. Lithium batteries use lithium metal as the anode. The voltage is 3V, which is twice as high as the zinc-carbon and alkaline manganese batteries. Lithium batteries have the advantage over the conventional zinc-carbon and alkaline manganese batteries in terms of voltage, energy capacity and the shelf life.

Sanyo of Japan developed lithium batteries which consist of the lithium anode and manganese dioxide cathode in combination with organic solvent in 1970s. These batteries were widely used in the automatic camera in the form of a 6V battery pack. Manganese dioxide used in lithium batteries was treated to completely remove the water since lithium reacts with water to form a hydrogen gas. Operation of lithium batteries was made possible by forming the protective film on the metal surface which prevents the lithium metal from reacting with the electrolyte. Lithium batteries can be dangerous under the abused condition. The abused condition of the primary batteries is rather simple. When the metallic object touches the positive and negative terminal simultaneously, there will be a short in the battery. In case that one of cells in the multi-cell connection is wrongly connected by mistake, this cell will be charged by other cells. In order to eliminate the possibility of the abuse condition, lithium batteries were supplied in the form of a pack when the multi-cell connection is required like a 6V automatic camera power pack.

Under the inspiration of the success of Sanyo's lithium primary batteries, Panasonic of Japan developed lithium batteries that differentiate its batteries from those of Sanyo. Panasonic's strategy was to use fluorine as the cathode material for the purpose of combining lithium metal anode with the most electro-negative material in the second period of the periodic table. However, the problem associated with the fluorine cathode material was that it is a gas at the ambient temperature. As an attempt to resolve this problem, Panasonic introduced the idea of intercalation of fluorine with the solid material. Based on the experimental results, carbon was selected as the host material for intercalation. As a result of numerous attempts, they found a way to intercalate fluorine atoms inside the carbon structure to make a solid compound, which is identified as (CF)n. These batteries were also used in the automatic camera in competition with

Sanyo's lithium batteries. As regards the cell performance of Panasonic's lithium primary batteries, they were similar to Sanyo's lithium batteries in the general performance and they were superior to Sanyo's lithium batteries in the shelf life after high temperature storage.

Lithium batteries have been in wide use in military applications because of their superior performance in spite of the fact that they can be intrinsically dangerous under the abused condition. Field incidents like fire or explosion take place when batteries are used by those who handle the batteries haphazardly. Soldiers are trained to comply with the rule and follow the protocol, so they are allowed to use high capacity lithium batteries in a real environment.

Cathode materials used in lithium batteries for military applications are SO_2 and $SOCl_2$. Unlike small lithium batteries mainly developed in Japan for commercial purpose, lithium batteries used in military applications were intensively developed by the USA. As an example, large lithium primary batteries were developed as a part of aerospace program in 1960s by NASA. The large $Li/SOCl_2$ battery used in auxiliary power system for missile weighs 68kg and contains the energy of 10,000Ah. Tadiran of Israel is also famous for lithium batteries. Israel is also strong in defense industry. By and large, batteries adopting the lithium metal anode attracted the attention of the military market where performance is the primary concern.

4) Secondary batteries

The secondary battery is the battery that can be reused by charging. We call the secondary batteries rechargeable batteries in contrast with the disposable batteries. Lead acid and nickel cadmium batteries which were developed in 1859 and 1899 respectively have been widely used in various

Batteries	Lead acid batteries	Nickel cadmium	Nickel metal hydride	Lithium ion batteries
Years	1859	1899	1990	1991
Inventors	Gaston Planté (France)	Waldemar Jungner (Sweden)	Toshiba Battery Panasonic	Sony
Voltage	2V	1.2V	1.2V	3.6V
Anodes	Lead (Pb)	Cadmium (Cd)	Metal hydride	Graphite
Cathodes	PbO_2	NiOOH	NiOOH	$LiCoO_2$
Electrolytes	H_2SO_4 solution	KOH solution	KOH solution	Organic solvent

Figure 2. The type of secondary batteries

applications for a long time. After the elapse of over ninety years, modern secondary batteries like nickel metal hydride and lithium ion batteries came into the picture.

Figure 2 shows the characteristics and development history of four major secondary batteries. Lead acid batteries have been used in the automobile as the starting, lighting & ignition (SLI) battery, and nickel cadmium batteries are widely used in the power tool application. Such modern batteries as nickel metal hydride and lithium ion batteries have led the battery industry since 1991. Their application area encompasses electric vehicles as well as portable electronics such as mobile phones, laptop computers and the camcorder.

(1) Lead acid batteries

Lead acid batteries were developed by French scientist, Gaston Planté, in 1859. They are still widely used in many applications. The main application of the lead acid batteries is the SLI battery of automobiles, industry vehicles like an electric forklift and the uninterruptible power supply (UPS). Lead acid batteries are also used in the power source of a submarine because of the amenability to make large batteries.

Lead and lead dioxide are used as the anode and cathode materials respectively, and an aqueous sulfuric acid solution is adopted as the electrolyte. The average voltage is 2V, but the charging voltage rises up to 2.5V, which is much higher than the voltage of electrolysis of water, 1.23V. However, the kinetic barrier inhibits the dissociation of water. The key feature of the battery reaction associated with lead acid batteries is the fact that the electrolyte is involved in the battery reaction. As a result, the concentration and the concomitant density of the electrolyte change with the battery reaction. Density change of the electrolyte, which has the harmful effect on the battery performance, is the major cause of degradation.

Listed below is the battery reaction of lead acid batteries:

The anode reaction: $Pb = Pb^{+2} + 2e^{-}$

$Pb^{+2} + SO_4^{-2} = PbSO_4$

The cathode reaction: $PbO_2 + 4H^{+} + 2e^{-} = Pb^{+2} + 2H_2O$

$Pb^{+2} + SO_4^{-2} = PbSO_4$

The battery reaction: $Pb + PbO_2 + 2H_2SO_4 = 2PbSO_4 + 2H_2O$ (2)

Participation of the electrolyte in the battery reaction brings about the non-uniformity of the electrolyte concentration. This problem was resolved by the introduction of a new separator, giving birth to the advanced lead acid batteries called absorbent glass mat (AGM) batteries. AGM batteries are considered as the major candidate for start-stop application (SSA) vehicles. SSA vehicles are also called Micro HEVs.

Anode and cathode plates are placed vertically inside the container which holds the electrolyte. The grid acts as both the current collector and the substrate of the electrode. As pure lead is too soft, alloying element is added to give the grid the strength that the battery requires. Antimony (Sb) is commonly used with the exception of several cases in which calcium (Ca) should be added to the lead instead of antimony. In the maintenance-free battery, the electrolyte consumption should be minimized. Use of calcium as the alloying element reduces the overcharging current, which minimizes the electrolyte consumption caused by overcharging. Another application area of the use of calcium as the alloying element of the lead is the submarine. In the enclosed area like a submarine, antimony is in danger of generating a harmful gas during charging, so calcium should be used to prevent the accidental evolution of a harmful gas.

Unlike other secondary batteries, the lead acid battery should be maintained under charged state as the battery is prone to be damaged at the over-discharged condition. Trickle charging is adopted to preserve the battery in the full charged condition. Trickle charging is the charging method in which we charge a fully charged battery under no-load at a rate equal to its self-discharge rate. Trickle charging makes the battery remain at its fully charged level.

When the lead acid battery is over-discharged, the interfacial resistance between the grid and the electrode increases due to the formation of the passive film. The passive film formed in the interface prevents the movement of electrons. This phenomenon is called the sulfation. Once the passive interfacial film is formed, it cannot be removed by electrochemical means, which results in the permanent damage of the batteries. When an automobile is lying idle for a long period of time, SLI batteries are over-discharged by a self-discharge phenomenon, which results in the degradation of batteries by sulfation.

(2) Nickel cadmium batteries

The nickel cadmium battery was invented by Waldemar Jungner of Sweden in 1899, and Varta of Germany and SAFT of France had played an important role in commercialization of nickel cadmium batteries. The prominent feature of nickel cadmium batteries in contrast with lead acid batteries is the high rate discharge capability. Nickel cadmium batteries employ alkaline solution as the electrolyte. Such alkaline solution as KOH solution has the highest ionic conductivity among electrolytes, so nickel cadmium batteries can generate the current high enough to drive the power tool. At first, a pocket type, in which active materials are filled inside the metal can with many small holes, was adopted. After World War II, a sintering type, in which the nickel substrate acts as both the current collector and substrate, was introduced to improve the cell performance. Interestingly enough, introduction of a sintering type electrode made nickel cadmium batteries subject to memory effect. In spite of the inconvenience caused by the memory effect, nickel cadmium batteries were in good use in many applications.

Cadmium and nickel oxyhydroxide are used as the anode and cathode materials respectively, and the KOH solution is adopted as the electrolyte.

The battery reaction of nickel cadmium batteries is as follow:

$$Cd + NiOOH = Cd(OH)_2 + Ni(OH)_2 \qquad (3)$$

The voltage of nickel cadmium batteries is 1.2V. Unlike lead acid batteries, nickel cadmium batteries exhibit a flat discharge curve.

In the batteries that use the aqueous electrolyte, the amount of the electrolyte decreases if they are exposed to the overcharging condition by the electrolysis of water. The electrolyte consumption by the repeated overcharging makes it necessary to add the electrolyte to the battery on a regular basis. The cost and inconvenience associated with the maintenance of the secondary battery led to the development of a maintenance free battery. The maintenance free battery is the battery that eliminates the need of the electrolyte addition. Oxygen generated in overcharging at the cathode is absorbed at the anode to prevent the dissociation of water, which is called the oxygen recombination mechanism. It plays a central role in the development of a maintenance free battery. Oxygen recombination mechanism led to the development of a maintenance free nickel cadmium battery. In fact, oxygen recombination mechanism enabled Sanyo of Japan to invent a small nickel cadmium battery used in portable electronics in 1960. Thanks to this invention, the application area of the secondary battery was expanded to the portable electronics, a new marketing horizon of the secondary battery.

Oxygen recombination mechanism is as follows. An oxygen gas is evolved in the cathode under overcharging condition by the electrolysis of water.

$$2OH^- = 1/2\ O_2 + H_2O + 2e^-$$

An oxygen gas is moved to the anode to react with the anode materials. During this process, an oxygen gas is absorbed by the anode material by the following reaction.

$$Cd + 1/2\ O_2 + H_2O = Cd(OH)_2$$
$$Cd(OH)_2 + 2e^- = Cd + 2OH^-$$

For the proper operation of the oxygen recombination mechanism, the energy capacity of the anode should be larger than that of the cathode.

The nickel cadmium battery has a balanced battery performance, which makes it very reliable in various applications. Some people argues that the nickel cadmium battery is very close to our definition of a universal battery. The universal battery is the battery that can be conveniently used in every possible applications competitively. Cadmium used as the anode material of the nickel cadmium battery is considered as the ideal battery material in that it induces fast battery reaction. However, the finding that the cadmium metal is the cause of the deadly disease called "itai-itai disease" made people have the second thought on the nickel cadmium battery. People were beginning to think that cadmium metal of nickel cadmium batteries should be replaced by a more innocuous material.

(3) Nickel metal hydride batteries

Although nickel cadmium batteries showed well-balanced cell performance as well as the excellent power capability, the cadmium metal used in the anode was found to be the cause of disease called "itai-itai disease". Itai means "it hurts" in Japanese. As the name of the disease implies, it is said to inflict an excruciating pain to the victims. Hence, the need arose to replace the health-hazard cadmium metal with a harmless material.

When the battery industry was searching for an alternative to the health-hazard cadmium metal, a new alloy which had a wide spectrum of applications emerged and became a center of attention in the industry. It was a hydrogen absorption alloy. Battery industry opted for a hydrogen absorption alloy as the candidate for the replacement of cadmium because of its electrical compatibility with the cadmium anode. This battery was called a nickel metal hydride battery. It was commercialized by Panasonic and Toshiba Battery in 1990. Since nickel metal hydride batteries have the excellent electrical compatibility with the nickel cadmium batteries in terms of the voltage and discharge curve profile, they have rapidly replaced nickel cadmium batteries.

A hydrogen storage capability increases from 200 to 800 to 1,000 times as the means of storage is changed from gas compression to liquefying of gas to the use of hydrogen absorption alloys. A hydrogen absorption alloy is a kind of catalyst, the action of which depends on the temperature and pressure. A hydrogen gas is dissociated into hydrogen atoms in contact with the hydrogen absorption alloy by the catalytic reaction. Hydrogen atoms pass into the crystal structure of the hydrogen absorption alloy to form the metal hydride. In case the need arises to recover a hydrogen gas, the temperature and pressure are to be changed to the recovery condition. Hydrogen atoms are moved to the surface of the hydrogen absorption alloy and form a hydrogen gas. The hydrogen storage mechanism of a hydrogen absorption alloy is described by the following reaction.

$$M + 1/2\ H_2 = MH + Q\,(\text{heat}) \qquad (4)$$

where M: metal, MH: metal hydride

Depending on the materials, equation (4) can be an either endothermic or exothermic reaction. Design methodology of a hydrogen absorption alloy is to combine endothermic and exothermic reaction materials to enable it to be activated at the proper temperature and pressure. Among various hydrogen absorption alloys, AB_2 and AB_5 alloys are commonly used in nickel metal hydride batteries.

Listed below is the example of hydrogen absorption alloys used in nickel metal hydride batteries.

AB_2: $ZnMn_2$, ZrV_2, $ZrNi_2$
AB_5: $CaNi_5$, $LaNi_5$, $NmNi_5$
where Nm: misch metal[3]

The battery reaction of nickel metal hydride batteries is as follows:

$$MH + NiOOH = M + Ni(OH)_2 \quad (5)$$

The nickel metal hydride battery had been the major battery in the portable electronics market in the first half of 1990s, but its market was encroached upon by lithium ion batteries in the second half of 1990s. High power nickel metal hydride batteries of D size (diameter 34.2mm, height 61.5mm) were adopted as the power source of the HEV by Toyota. In effect, the nickel metal hydride battery was the first battery that was applied to the HEV.

3 Nm represents rare earth metal mixture such as La, Ce.

03
:
Lithium secondary batteries

A remarkable increase in the interest of lithium secondary batteries occurred in the 1980s probably inspired by the commercial success of lithium primary batteries in the 1970s. As a way to rekindle the worldwide interest in the lithium batteries, battery scientists and engineers in Canada gathered together to establish Moli Energy for the purpose of commercializing lithium secondary batteries incorporated with liquid electrolytes. In Europe, a polymer electrolyte, which acts as both a separator and the electrolyte simultaneously, was utilized to develop the LPB. On the other hand, Japan was keen to find the way to intercalate lithium atoms to the graphite for the development of lithium ion batteries. Competition between lithium secondary batteries and LPB and lithium ion batteries was intense in the 1980s. In this respect, the 1980s was arguably considered as the decade of the competition of the development of lithium rechargeable batteries.

1) Moli Energy's lithium batteries

Moli Energy developed lithium secondary batteries with its emphasis placed upon the improvement of the soundness of the solid electrolyte inter-phase (SEI) film formed on the lithium metal surface. The SEI film is an ion conducting film that inhibits reaction of lithium metal with the electrolyte. It was found that the quality of a SEI film depends on the electrolyte and a lithium salt. Through the comprehensive evaluation procedure, Moli Energy decided the battery components that fit into the target cell performance and the safety requirement. Lithium metal foil and molybdenum disulfide were selected as the anode and cathode materials. The electrolyte was the solution of propylene carbonate (PC) and ethylene carbonate (EC) with the addition of 1M of $LiAsF_6$. Moli Energy called its batteries MOLICELL™ and commercialized lithium secondary batteries in 1988.

Superior performance of lithium secondary batteries of Moli Energy attracted attention of the battery industry. NEC, a mobile phone maker of Japan, adopted Moli Energy's lithium secondary batteries in its mobile phones. However, in the midst of use of NEC's mobile phones, there occurred the incident in one of their mobile phones in Japan. Fire broke out in the power pack of lithium secondary batteries. Gas was built up inside the can and the vent was opened to release the internal pressure of the battery. Gas was discharged through the open vent, and in this process flame broke out through the vent. Mobile phone industry including NEC was shocked by this field incident because at that time fire or explosion of the mobile phones was absolutely unthinkable. NEC decided to recall its mobile phones powered by lithium secondary batteries in August, 1989 to prevent the recurrence of the field incident. In this turmoil, Moli Energy was merged to NEC, and the development activities of the lithium secondary batteries had been dwindled since then. Instead, nickel metal hydride and lithium ion batteries were given the chance to enter the portable electronics market.

2) Problems and remedies

It would be interesting to investigate the underlying cause of the demise of lithium secondary batteries incorporated with liquid electrolytes. Vulnerability of lithium secondary batteries to the field incidents like fire or explosion is related to the combination of lithium metal with liquid electrolytes.

The solid electrolyte inter-phase (SEI) film is formed by the reaction of lithium metal with liquid electrolytes. This ion conducting film hinders lithium metal from reacting with liquid electrolytes. As long as the soundness of the SEI film is maintained, safety of lithium secondary batteries is secured. However, the SEI film is prone to be damaged by the non-uniform flow of lithium ions. When lithium ion flows are concentrated on the specific location of the SEI film, the film is damaged locally, which aggravates the uniformity of lithium ion flows further. Local destruction of the SEI film increases the chance of either dendrite growth on the lithium metal surface or electrolyte penetration into the lithium metal, which increases the likelihood of field incidents like the outbreak of fire or explosion.

Degradation mechanism of the lithium metal anode implies that removal of either lithium metal or liquid electrolytes would insure the safety of lithium secondary batteries. Therefore, the way to use lithium secondary batteries safely is to remove the possibility of the combination of lithium metal with liquid electrolytes. A polymer electrolyte is the polymer film that acts as both the electrolyte and a separator simultaneously. Replacement of the liquid electrolyte with a polymer electrolyte film produces a LPB, which makes lithium batteries safe by the elimination of liquid electrolytes. Substitution of a lithium metal anode with carbon or graphite makes the lithium ion battery, which improves the safety of lithium batteries by the complete removal of lithium metal. Therefore, both lithium polymer batteries and lithium ion batteries manifested themselves as a means to rectify the problems associated with Moli Energy's lithium secondary batteries.

3) Lithium polymer batteries

(1) EC project

In the early 1980s European Community funded a LPB project to further develop lithium secondary batteries incorporated with a polymer electrolyte. These were French-Canadian and Anglo-Danish project.

Elf, French oil company, collaborated with Hydro-Québec, Canadian electric power company, to further develop the LPB technology based on its basic patents.[1] Atomic Energy Association (AEA), Research Institute of the UK, formed collaborative research team with Innovision, Danish Research Institute, to put LPB technology one step closer to commercialization. The former was called a French-Canadian project and the latter an Anglo-Danish project. The emphasis of a French-Canadian project was on the improvement of the dry polymer. On the other hand, the purpose of an Anglo-Danish project was to develop an ambient temperature LPB by extending operating temperature range. The Anglo-Danish project chose polyethylene oxide (PEO) as the base material for a polymer electrolyte and added lithium perchlorate ($LiClO_4$) as the lithium salt. As a measure to extend the operating temperature range, plasticizers were added to the polymer electrolyte in combination with cross-linking of polymers. A polymer electrolyte film was laminated with lithium

1 M. B. Armand, J. M. Chabagno, and M. Duclot, in *Second International Meeting on Solid Electrolytes, St Andrews*, Scotland, 20-22 Sept. 1978.

metal foil anode and the cathode film to form a LPB. A cathode film is composed of vanadium oxide active materials, polymer electrolytes and carbon black. Inclusion of polymer electrolytes in the cathode is necessary to form the electrolyte channel in the cathode. Battery components and cell design along with the manufacturing methodology of an Anglo-Danish project had provided the foundation of the commercialization of LPB. Valence Technology of the USA was keen to utilize the results of an Anglo-Danish project for the commercialization of LPB ahead of other advanced batteries.

(2) Commercialization activities

Valence Technology located in San Jose, California was well-known to the battery industry in the early 1990s because of its potentials to compete with Sony of Japan as to which batteries would dominate the market in the future. While Sony was strong in inorganic chemistry, Valence Technology possessed strong technological background in the area of organic chemistry required to conduct the LPB development. A polymer electrolyte consisting of PEO and lithium perchlorate shows poor ionic conductivity at the ambient temperature, so plasticizers are added to extend the operating temperature range. However, the addition of plasticizers increases the interfacial resistance between PEO and plasticizers. Valence Technology's strategy was to minimize the interfacial resistance to form a monolithic structure devoid of any interface. Another strategy of Valence Technology was to reduce the manufacturing cost by the design of a continuous production equipment. A conventional manufacturing method of LPB was to make the cathode and a polymer electrolyte film separately and put them together with lithium foil by hot lamination process. The problem associated with the conventional production method is the possibility of the formation of air pockets in the interface of films. For the efficient elimination of the possibility of air pockets, a polymer electrolyte was coated directly onto the surface of the cathode film. In addition, cross linking method by electron beam was adopted to immobilize plasticizers.

With regard to the assembly process of pouch batteries, Valence Technology made a contribution to the fabrication of the electrode pocket and the zig-zag folding of electrodes. As compared with the prismatic battery with a metal can casing, it was hard to increase the thickness of the battery in the pouch battery. As a way to make a pouch case with a sizable thickness and robustness, Valence Technology developed the fabrication method of a pouch case by pressing the aluminum sealing paper on the fabrication die. The zig-zag folding method came from its efforts to automate the battery assembly process. Both sides of the lithium foil were used as the anode, which is referred to as the bi-cell configuration. The cathode and a polymer electrolyte film are folded like an accordion, and the rectangular lithium metal foils are inserted alternately. Since the zig-zag folding method is amenable to automation, it has been widely used in lithium ion pouch batteries.

Motorola, one of the leading mobile phone makers in the 1990s, collaborated with Valence Technology to develop LPB for the mobile phones. Valence Technology made LPB samples in its pilot plant and tested them in Motorola's mobile phones, but unfortunately LPB made in Valence Technology failed to deliver enough power to operate the mobile phones. Limited cycle life was another problem of LPB which Valence Technology had presented to Motorola as testing samples. Valence Technology's failure to meet the requirement of mobile phones made people have the second thought on the prospect of LPB, which was reflected in the stock price of Valence Technology immediately. Valence Technology was forced to move to Henderson, Nevada from San Jose,

California because of the financial difficulties. In this process, many battery engineers left Valence Technology. To maintain the viability within the battery industry, Valence Technology switched the target battery from LPB to plastic lithium ion (PLI) batteries of Bellcore.

Bellcore of the USA had developed lithium ion batteries adopting spinel manganese oxide (LMO: $LiMn_2O_4$) cathode material in the early 1990s. In the midst of the development of lithium ion batteries, they were informed that Sony commercialized lithium ion batteries successfully. Deprived of the driving force for the development of lithium ion batteries, Bellcore put a stop to the lithium ion battery development and decided to switch to the solid form of the lithium ion battery, which was called the PLI battery. Bellcore made the solid form of lithium ion batteries by using a polyvinylidene fluoride (PVDF) copolymer. Its target was to make batteries with monolithic structure by reducing the interfacial resistance between electrodes and a polymer electrolyte.

The polymer film Bellcore developed was not exactly a polymer electrolyte since the liquid electrolyte impregnated in the polymer film acted as the medium of the movement of lithium ions. Because of the introduction of the liquid electrolyte in the polymer matrix whose function is merely to hold the electrolyte, the PLI battery exhibited high power capabilities and good low temperature performance which had been considered as the main problem of batteries incorporated with a polymer electrolyte. Working mechanism of PLI batteries is same as that of lithium ion batteries except that PLI batteries is bulkier than lithium ion batteries.

After Valence Technology's adoption of PLI battery technology, it designed the battery packs to compete with cylindrical lithium ion batteries in the mobile phone market. In the mid 1990s, the operating voltage of mobile phones was 5V, so two cylindrical lithium ion cells were connected in series to produce a 7.2V pack. Because of the use of cylindrical batteries, some portion of the dead space in the pack was unavoidable. Valence Technology decided to utilize a favorable form factor of PLI batteries. It designed the power pack with higher energy capacity by removing the dead space. Unfortunately, its well-thought battery design plan was thwarted by the introduction of 3V mobile phones. Motorola developed a 3V power amplifier of the mobile phone, which was the last remaining 5V component among the mobile phone components. Introduction of a 3V mobile phone made it possible to design a pack with a single battery. If a single battery is to be used in the mobile phone, the prismatic lithium ion battery is the best choice. In actuality, prismatic lithium ion batteries dominated the mobile phone market after the change of the operating voltage of the mobile phone.

In order to hold the same amount of electrolytes as lithium ion batteries, the volume of PLI batteries should be increased. As a result, energy density of PLI batteries was intrinsically lower than that of prismatic lithium ion batteries, which means that PLI batteries take up more space than prismatic lithium ion batteries. At that time, mobile phone market preferred small and light batteries with a slim design. The PLI battery was not able to compete with the prismatic lithium ion battery in the mobile phone market mainly because of the low energy density.

In conclusion, the PLI battery was outperformed by the lithium ion battery in the mobile phone market. The market PLI batteries could have penetrated into dwindled significantly.

(3) Lithium-sulfur batteries

In the 1990s, there was an effort to make 2V lithium secondary batteries by combining lithium metal anode with the sulfur cathode. Operating voltage of mobile phones was reduced from 5V to 3V. People thought that the 2V mobile phone would come out in a perceivable future. To meet the requirement of the future mobile phone, lithium-sulfur (Li-S) batteries were being developed. Lithium-sulfur batteries enhance the safety associated with a lithium metal anode, which is called the shuttle cock mechanism. Shuttle cock mechanism removes the dendrites and dendrite embryos on the lithium metal surface by polishing effect. Sulfur compound is formed by the reaction of the sulfur cathode with the electrolyte, which acts as the polishing agent. These polishing agents are transported to the lithium metal anode and remove the dendrites and embryos of dendrites. After removal of the dendrites, the polishing agents move back to the cathode and are absorbed to the cathode.

Moltech located in Tuscon, Arizona and Polyplus, a start-up company of Lawrence Berkeley Laboratory, developed lithium-sulfur batteries in 1990s. Moltech aimed at commercialization of lithium-sulfur batteries whereas Polyplus put more emphasis on the fundamental research. Monsanto, a global chemical company in the USA, provided five million dollars research fund and Ericsson of Sweden, a major mobile phone company in the 1990s, furnished three million dollars for the early adoption of Moltech's lithium-sulfur batteries. Nokia of Finland, the leading mobile phone maker of the 1990s, requested lithium-sulfur battery samples for testing. Moltech asked Lehmann Brothers, the investment bank of the USA, to find the battery firm which can commercialize lithium-sulfur batteries using its technology. Lehmann Brothers made contact with LG Chemical, a Korean battery firm, through M&A team of LG Security in 1997. Discussion between Moltech and LG Chemical in relation to the technology transfer of lithium-sulfur batteries continued on and off for two years without any tangible results.

During the period of discussion between Moltech and LG Chemical,

Sony and Hitachi-Maxell of Japan tested Moltech's lithium-sulfur battery samples, but contrary to their expectation, the result was disastrous because of the swelling of cells. The news on the test result of Moltech's samples was passed on to LG Chemical by a one man battery consulting firm in Japan. Hearing the news on the sample test results, LG Chemical decided to drop the lithium-sulfur project in 1999. In contrast, Samsung SDI of Korea had made equity investment to Polyplus before the problem associated with swelling of cells was spread in the battery industry.

Swelling of lithium-sulfur cells is found to be caused by the shuttle cock mechanism which makes the lithium-sulfur battery safe. In lithium-sulfur batteries, the polishing agent is made by dissolution of the cathode material in the electrolyte. In other words, chemical reaction takes place in lithium-sulfur batteries, which means that lithium-sulfur batteries violate the basic principle of electrochemical reaction, in which no electron exchange should take place between the battery components. After removal of dendrites and dendrite embryos, some of the compounds fail to go back to the cathode. Instead, they move around the electrolyte and accumulate inside the battery, which results in severe swelling of the battery. Swelling in the battery is a very serious problem in that it can distort or break down the plastic power pack case, so it is an irreconcilable defect.

4) Lithium ion batteries

The cathode materials of lithium secondary batteries exhibit the excellent performance in terms of cycle life. However, the combination of the cathode materials with the lithium metal anode shows the premature demise of the cycle life. Acceptance of this finding would imply that we may be able to design the high cycle life battery by replacing the lithium metal anode by the materials with crystal structure similar to the cathode. This concept was the driving force for the development of a swing battery.

In the swing battery, there is no supply source of lithium atoms, and lithium atoms move to and fro between the anode and cathode during charge and discharge. That's why it is also called a rocking chair concept battery. Sony called the swing batteries lithium ion batteries to stress that swing batteries are safe because lithium exists in the form of ions instead of atoms. The most promising candidate for the anode material was graphite as the voltage is similar to that of lithium metal and the energy capacity is high. The specific capacity of graphite is 372mAh/g. In graphite, six carbon atoms with the shape of hexagon are intercalated with one atom of lithium, forming LiC_6. However, as graphite is intercalated with the residue of propylene carbonate (PC) electrolyte molecules, cycle life drops drastically in the initial stage. This is because of the co-intercalation of PC electrolyte molecules, but the reason for the premature demise of the cycle life was not completely understood at that time.

In 1983, Japanese scientists succeeded in intercalation of lithium atoms with graphite in the polymer electrolyte. In the second half of the 1980s, Japanese battery makers changed the anode from graphite to carbon and intercalated lithium atoms with carbon in the liquid electrolyte environment successfully. Thanks to the success of intercalation of lithium atoms with carbon, the basic technology associated with lithium ion batteries was finally established. The next step was to make these batteries working commercially. Sony came forward to do this historical job.

PART

2

LITHIUM ION BATTERIES

04
:
Development of lithium ion batteries

1) Sony's commercialization

(1) The target battery

In the 1980s, Sony was famous for its ingenious inventions in the area of portable electronics such as a Walkman and a camcorder. It is said that the founder of Sony, who majored in music in the university, named his company Sony based on the term Young Sound. Sony emphasized the creativity, and its strong emphasis on the creative activities made it possible to develop new portable electronics ahead of the competitors. It is reported that Sony didn't allow collaboration with other companies on the ground that it would hurt creativity of Sony.

With regard to the battery business, Sony had been making AAA alkaline manganese batteries by the acquisition of alkaline manganese battery technology from Union Carbide of the USA. Sony used AAA primary batteries in the remote TV controller. Therefore, although Sony was not considered as the battery company in the 1980s, it had some knowledge and experience pertinent to the battery business. In the late 1980s, Sony searched for secondary batteries for the camcorder, for the conventional nickel cadmium battery was not satisfactory in terms of energy capacity and cycle life. Sony wanted to acquire batteries that could outlive the camcorder lest camcorder users should replace batteries during the lifetime of the camcorder. Meanwhile, Japanese chemical company called Asahi Kasai developed a swing battery incorporated with carbon anode through the success in the intercalation of lithium atoms to the carbon anode. However, Asahi Kasai decided not to carry forward the commercialization of the swing battery because Asahi Kasai's senior managers felt that it would be unreasonable for the chemical company like Asahi Kasai to try to compete with big battery companies like Sanyo and Panasonic. They may have thought that Asahi Kasai didn't have the capability to form the sales channel from the ground up in the highly competitive battery market.

After Asahi Kasai's strategic decision not to commercialize the swing battery, its invention had been handed over to Sony. It is said that one of Asahi Kasai's battery scientists gave the samples of swing batteries to his friend in Sony as his friend showed interest in Asahi Kasai's batteries in the private gathering. After extensive investigation on the possibility of the swing battery through handed-over battery samples, Sony decided to go ahead with further development of the swing battery by compensating for the limitations of Asahi Kasai's batteries. Sony called the swing batteries lithium ion batteries to stress that swing batteries are safe because lithium exists in the form of ions instead of atoms. This is how lithium ion batteries and Sony came into the picture in the battery industry.

After Sony opted for a lithium ion battery as the target product, battery engineers of Sony designed the lithium ion battery and determined the cell formula and manufacturing recipe. They adopted hard carbon and lithium cobalt oxide as the anode and cathode materials respectively. In the beginning, they developed 14500 AA (diameter 14mm, height 50mm) and 20500 (diameter 20mm, height 50mm) size cylindrical batteries because these cylindrical batteries had been commonly used in the battery industry. The AA size is the most common alkaline manganese battery, so the choice of this size exhibits Sony's intention to make lithium ion batteries compatible with alkaline manganese batteries in terms of the battery size. The 20500 size is the popular size in the nickel cadmium and nickel metal hydride batteries. The choice of 20500 size reflects Sony's desire to replace the conventional secondary batteries with lithium ion batteries. Sony's lithium ion batteries were installed in the mobile phones in June, 1991, which is considered as the year of commercialization of lithium ion batteries.

However, 14500 and 20500 size batteries did not fit into the requirement of the camcorder. Sony's battery engineers wanted to design the battery that fits right in with their camcorder, and with this in mind, they designed the battery with a right size for the camcorder based on the market information. The market survey indicated that the main users of Sony's camcorder were middle-aged oriental men. The 18650 size, whose diameter is 18mm and height 65mm, was determined as the target battery size not only because it would give enough energy capacity to the camcorder but also because this size would give main users, middle-aged oriental men, great comfort when they grabbed the battery. Energy capacity of an 18650 cylindrical batteries was 1.2Ah at that time, which was same as that of 20500 cylindrical batteries. Although it is the specially designed size for Sony's camcorder, an 18650 cylindrical batteries became the most commonly used lithium ion battery with a wide spectrum of applications.

(2) Cathode materials

Lithium cobalt oxide (LCO: $LiCoO_2$) became the major cathode material in the 1990s since Sony employed the LCO cathode in the cylindrical batteries in 1991. Interestingly enough, Sony's choice of LCO over other cathode materials was not based on the investigation on the pros and cons of each material. Rather, this choice stemmed from the experience and knowledge on cobalt metal acquired during the time when Sony was developing samarium cobalt (Sm-Co) permanent magnet. LCO is easy to manufacture because it is made by simple solid state reaction, and has a simple primary structure in powder morphology. These features have the advantage of giving reliable and consistent electrical properties regardless of the production method. Taking these favorable characteristics of LCO into consideration, Sony's choice of LCO as the cathode active material turned out to be an excellent one. Otherwise, Sony wouldn't have succeeded in commercialization of lithium ion batteries.

The most well-known patent in the arena of lithium ion batteries is that of LCO. This patent is concerned with 4V cathode materials with layer structure, and it belongs to Atomic Energy Association (AEA) of the UK. Professor Goodenough of University of Texas, Austin went to University of Oxford for sabbatical leave in the late 1970s and stayed in AEA located close to University of Oxford for the battery materials research since AEA had a good battery research foundation. He developed cathode materials at AEA, the fund being provided by Dowty, a lithium battery company close to AEA.

LCO has the layer structure. Oxygen atoms, the largest atom among three elements, form the frame of the layer structure, and the cobalt atoms layer is inserted between two oxygen layers, which provides the basic structure unit like a piece of bread in the sandwich. Lithium atoms which have high mobility electrochemically by the mechanism called the intercalation form the lithium atoms layer between two structure units consisting of cobalt and oxygen. The patent of LCO covers all the cathode materials which have the layer structure including NCA and NCM. In spite of the successful development of 4V cathode materials, LCO didn't attract the attention of battery industry at that time since battery engineers thought that 4V batteries could not be feasible in the age of 1-2V batteries. As a result, it was practically buried in oblivion.

In 1994, Dowty of the UK was sold to Panasonic of Japan on account of its continuous financial difficulties, and patents which Dowty possessed were supposed to change hands. In the situation like this, Sony paid a visit to AEA and proposed an enticing deal to AEA. Sony's proposal was as follows. If AEA prevents the LCO patent from going on to Panasonic, Sony will pay the patent fee to AEA and share the subsequent licensing fee and royalty accruing from licensing equally. AEA was flabbergasted by Sony's offer because AEA didn't realize the gravity of a LCO patent at that time. In accordance with Sony's request, AEA called a meeting to Panasonic, and proposed to give other AEA's patents in exchange for the LCO patent, on which Panasonic agreed. At that time, Panasonic thought that the nickel metal hydride battery would have a better future than the lithium ion battery. Panasonic's agreement on AEA's offer was a harbinger for Sony's successful lithium ion battery business. If Panasonic had said no to AEA's offer, Sony would have had to report to Panasonic the business performance on a regular basis. If that had happened, we cannot rule out the possibility that Sony, a creativity driven company, would have given up on the lithium ion battery business. After AEA's resolution of a cathode patent, the LCO patent was owned by AEA and Sony jointly.

Since the early 1990s, British government had pushed hard to privatize public enterprises. As a part of this drive, AEA became AEA Technology. To run a newly converted commercial company, AEA Technology was in need of a source of revenues because British government no longer supported AEA Technology financially. The LCO patent played an important role in supporting AEA Technology both financially and morally. AEA and Sony received 1.2 million pounds for licensing fee and 2.5% running royalty from Japanese lithium ion battery makers and LG Chemical of Korea. Samsung SDI and BYD were exempt from the licensing fee and royalty because their sales of batteries took place after expiration of the patent.

Thanks to the influence of Sony, AEA Technology was able to have a much larger say and role in the lithium ion battery industry. Dr. Goodenough, the inventor of LCO, continued to be active in the battery materials research and developed another important cathode materials, lithium iron phosphate (LFP: $LiFePO_4$) in 1996. He received the Japan Prize in 2001 for his discovery of LCO.

(3) Protection circuit module

Senior managers of Asahi Kasai decided not to commercialize lithium ion batteries not only because Asahi Kasai didn't have the battery business background but also because they thought lithium ion batteries would not be able to function properly in a real environment. Although lithium ion batteries showed good battery performance under strictly controlled condition, they became unstable under overcharge, over-discharge and over-current condition.

Lithium ion batteries are charged in a mixed mode. They are charged at constant current until the battery voltage reaches 4.2V. The charging voltage was increased to 4.35V later on. After the constant current charging, charging mode is switched to a constant voltage mode. In constant voltage charging mode, lithium atoms are penetrated into the crystal structure of the graphite anode by field-affected diffusion until most of the available interstitial sites of the graphite structure are occupied. It will take about 3 hours to complete the charging. Lithium atoms in the lithium layers of the crystal structure of the cathode material are moved out of the cathode during charging. To preserve the layer structure of the cathode material, a certain amount of lithium atoms should remain in the lithium atoms layers. Overcharging is the situation in which excessive lithium atoms are moved out of the cathode. If that happens, the cobalt atoms in the cobalt layer tend to move to the lithium atoms layer, the crystal structure being transformed from the layer to a more stable cubic structure. In the crystal transformation process, excessive oxygen atoms are spewed out to accommodate the cubic structure. If the electrolyte is hot enough to reach the flash point, it catches on fire by reacting with oxygen spewed out from the cathode. Therefore, overcharging can bring about fire or explosion because lithium ion batteries have no overcharging protection mechanism. In other secondary batteries such as lead-acid, nickel cadmium and nickel metal hydride batteries, oxygen recombination mechanism prevents the battery from increasing the voltage beyond the charging voltage. Oxygen formed in the cathode in overcharging is moved to the anode and absorbed, which phenomenon is called the oxygen recombination mechanism.

Discharge termination voltage of lithium ion batteries is 3V. Lithium atoms in the graphite anode are moved out of the graphite anode structure during discharging. The extreme case of over-discharging is the state in which electrons are generated at the anode without the presence of lithium atoms inside the graphite. If that happens, copper foil used as the anode current collector is oxidized, and copper ions are moved to the electrolyte. Dissolution of copper foil makes the battery vulnerable to the short as copper is the metallic impurities. Therefore, over-discharged lithium ion batteries can catch fire because of the short. Like lead-acid batteries which suffer from sulfation phenomenon under over-discharging condition, lithium ion batteries are damaged and their performance is degraded by over-discharging.

High current discharge should also be avoided since it increases the battery temperature uncontrollably. One of the most serious situation in lithium ion batteries is the outbreak of thermal runaway, which can result from high current discharge. Therefore, it was necessary to keep the current low until sufficient information and pertinent data that can give us the confidence on the safety are accumulated. In conclusion, unlike other batteries, the lithium ion battery becomes unstable under overcharge, over-discharge and over-current condition. Likelihood of catching fire or explosion increases under these conditions because of the absence of the intrinsic control system.

In order to reduce the vulnerability of the battery to safety issues, it is required that the current be switched off when the voltage or current is out of range, so Sony designed the electronic circuit module which cuts off the current whenever the voltage or current is out of range. Sony referred to this module as the protection circuit module (PCM), which consisted of a Sony's proprietary microprocessor and transistor switches. To compensate for the absence of the internal control system, Sony had embedded the artificial control system in the battery. As a result of connecting the PCM to the lithium ion cells, lithium ion batteries could be operated as normally as the other batteries, but there was a hurdle to justify the use of the electronic circuit module. The battery is the power source, the function of which is to provide the electricity to the electronic circuit module. In this respect, it was hard to accept the fact that the battery that drives the electronic circuit module needs an electronic circuit for the proper powering of the target electronic circuit. In the beginning, the battery industry was reluctant to accept lithium ion batteries on account of the need for PCM. Moreover, the price of lithium ion batteries at that time was unimaginably high. The price of a 2 cell power pack for Sony's camcorder was over 100 dollars. In actuality, it took some time until battery users learned to get accustomed to this awkward concept of the PCM and incredibly high price.

By the introduction of a PCM, lithium ion batteries could be used in a real environment like the other batteries. By and large, Sony's achievement is to make lithium ion batteries working in a real environment by taking advantage of the knowledge on electronic technology. When lithium ion batteries first appeared in the market in 1991, the battery industry was surprised at the enormously high cycle life. Cycle life was so high that lithium ion batteries could easily outlive the camcorder, mobile phones and laptop computers. Finally, the battery industry succeeded in inventing batteries that could outlive portable electronics, which eliminated the need for the battery replacement during usage of the portable electronics. Sony finally commercialized the battery we had dreamed of from the perspective of the battery life.

2) Characteristics of lithium ion batteries

(1) Definition

The lithium ion battery has the characteristics that differentiate itself from other batteries. In general, the lithium ion battery can be defined by the following three features.

(1) The lithium ion battery is the battery that has the graphite anode instead of lithium metal.
(2) The lithium ion battery is the battery that has the organic electrolyte whose reaction window is over 4V.
(3) The lithium ion battery is the battery that has the aluminum current collector of cathode.

First, the lithium ion battery is the one that has the graphite anode. The anode is the electrode in which oxidation reaction occurs during discharge. Every now and then, the anode is called the fuel electrode since it supplies electrons generated in the oxidation reaction to the outside, so metal is used as anode materials because metal is abundant with free electrons. Zinc of dry cell and alkaline manganese batteries, lithium of lithium primary batteries, lead of lead acid batteries, cadmium of nickel cadmium batteries and hydrogen absorbent alloy of nickel metal hydride batteries are examples of the metal anode. As regards the lithium metal anode, it has been used as the anode material for lithium primary batteries, but the effort to use lithium metal in the lithium secondary battery was not successful mainly due to the formation of dendrites on the lithium metal surface. One of the ways to circumvent the problem of the formation of dendrites on the surface of lithium metal is to replace lithium metal with graphite. That's why graphite was adopted as the anode material. Since graphite is used as the anode material, a lithium ion battery is the one that does not have the lithium supply source. Lithium ions move to and fro between the anode and cathode during charging and discharging, so people call these batteries swing batteries or rocking chair concept batteries. Problem in association with the swing type movement of lithium ions is that the amount of lithium ions can be reduced as a result of repeated cycles of charging and discharging. In order to prevent the occurrence of this phenomenon, it is required that movement of lithium ions be controlled with great accuracy. A PCM which was originally developed for the safety reason serves this purpose as well. It should be noted that the role of a PCM is not only to insure the safety of the battery but to maintain the energy capacity with cycles. Therefore, it is appropriate to consider a PCM as part of the lithium ion battery. Haphazard removal of the PCM might lead to premature demise of the cell performance. Since the use of a graphite anode makes the lithium ion battery so unique, it can be considered as the primary feature of the lithium ion battery.

Second, the lithium ion battery is the battery that uses the organic electrolyte whose reaction window is over 4V. When 4V cathode active materials like LCO were developed in the late 1970s, the battery industry didn't take interest in this cathode material because it thought that development of the 4V electrolyte was highly unlikely, although not impossible. However, development of organic solvent with reaction window of over 4V in the 1980s made it possible to complete the battery system. Most of materials disintegrate if the voltage exceeds 4V. That's why it was so difficult to find the electrolyte that does not disintegrate over 4V. Development of the electrolyte with the reaction window of over 4V accelerated the development of a high voltage lithium ion battery system. Unlike other batteries using aqueous electrolytes, lithium ion batteries use organic electrolytes because of the need of the high reaction window. Hence, the use of the organic electrolyte with the reaction window of over 4V is one of the key features of lithium ion batteries.

Third, the lithium ion battery is the battery that uses aluminum foil as the cathode current collector. Unlike other secondary batteries, charging voltage exceeds 4V. Most metals tend to disintegrate at the voltage of over 4V. Aluminum is one of a few metals that can withstand oxidation under this severe electrochemical condition. Therefore, use of the aluminum current collector is considered as one of key features pertinent to lithium ion batteries. Many people thought that the use of aluminum foil as the cathode current collector was too simple to be accepted as the key feature of lithium ion batteries. But the importance of this patent comes from the fact that without the choice of aluminum foil, lithium ion batteries could not have been in the market.

In sum, combination of the graphite anode, the organic electrolyte and the aluminum current collector gave birth to lithium ion batteries. If it had not been for any one of them, lithium ion batteries could not have existed. Looking at these three features closely for the common denominator, you'll find that they are all associated with materials in contrast with other secondary batteries. That's why some people call the lithium ion battery a materials-dominant battery. From this point of view, participation of such traditional chemical companies as LG Chemical and SK innovation in the lithium ion battery business is rightfully justified. The need for the materials technology in lithium ion batteries complicated and diversified battery technology. Japan is strong in materials technology, which explains why Japan surpassed western countries in the competition of lithium ion batteries. Japan's strength in lithium ion batteries comes from its capability to generate and analyze data systematically, which plays an important role in the development and improvement of materials-dominant batteries like the lithium ion battery.

(2) The battery design

We start with 18650 cylindrical batteries commercialized by Sony in 1992 for the investigation of the lithium ion battery design. This battery was targeted for use in Sony's camcorder. The energy capacity of this cell was 1.2Ah and the average voltage was 3.6V, charging voltage 4.2V and discharging cut-off voltage 2.7V. The prominent feature of the lithium ion battery is its high voltage, which is three times higher than the nickel cadmium battery. Since high energy capacity of the lithium ion battery stems from its voltage, any development activities that result in the decrease of voltage are considered undesirable. Sony used ring type propylene carbonate (PC) as the base solvent of the electrolyte, and added chain type dimethoxy ethylene (DME) to propylene carbonate for the control of viscosity and other pertinent properties. Ring type solvent, which is a polar solvent, is effective in moving lithium ions. The lithium salt which Sony had chosen among many candidate materials was $LiPF_6$. Sony's preference of $LiPF_6$ to the well-known $LiClO_4$ is based on the investigation results that $LiClO_4$ is not compatible with the lithium ion batteries in terms of safety. $LiClO_4$ salt is prone to grow dendrites on the anode surface under overcharging condition whereas $LiPF_6$ grows globular type lithium metals on the carbon anode surface in overcharging. Therefore, $LiClO_4$ is more vulnerable to internal shorts caused by the separator penetration. Furthermore, $LiClO_4$ increases the danger of fire during welding process.

Sony adopted hard carbon as the anode material instead of commonly used soft carbon by virtue of high energy capacity of hard carbon. Graphite is made from soft carbon by graphitization process whereas hard carbon cannot be transformed into graphite by heat treatment, which is due to the more irregular structure of hard carbon. The irregular crystal structure of hard carbon increases the energy capacity because of the more space for lithium intercalation.

LCO was used as the cathode material, and its manufacturing process was developed by Sony. Sony's battery engineers were able to develop the manufacturing technology of LCO since they knew how to deal with cobalt through their experience in the development of samarium cobalt (Sm-Co) permanent magnet for the speakers. The average particle size of LCO was 20 micrometers. They discarded particles less than 5 micrometers and over 40 micrometers. They removed oversized particles due to the danger of the puncture of a separator. Undersized particles not only make the mixing process difficult but also polarize the cathode electrode in charging and discharging because of the size difference. As regarding the particle size of the anode material, they used carbon which has double peak distribution of particle size not only to avoid point contacts between active materials but to increase the compactness of the electrode. Since electron conducting materials such as carbon black serves this purpose in the cathode, they used LCO powder with single peak distribution.

Electronic companies like Sony are traditionally weak in organic chemistry, so Sony relied on the knowledge and experience of the outside companies that specialized in organic chemistry to develop the battery materials related to the organic chemistry. Sony developed a separator in collaboration with Ube of Japan. It used 25 micrometers thick separators made by dry process. The separator made by dry process consists of three layers. Polyethylene film is sandwiched between two polypropylene films. When temperature reaches 140°C, a polyethylene film starts to melt, which leads to clogging of pores of polyethylene, the current being cut off by the blockage of the movement of lithium ions. This is called the chemical shutter of the separator. Chemical shutter plays an important role in preventing thermal runaway in the abused condition. Another safety device that hinders thermal runaway is the sophisticated vent called the current interrupt device (CID). It cuts off the current under overcharging condition. For the proper operation of this safety device, it is necessary to put a small amount of lithium carbonate (Li_2CO_3) in the cathode, which acts as the supply source of carbon dioxide. When the voltage reaches 4.7V, lithium carbonate starts to disintegrate and generate carbon dioxide gas, which increases the pressure inside the battery in proportion to the overcharging voltage. When the voltage reaches 5V, the pressure inside the battery becomes high enough to tear off the aluminum tab connecting the electrode to the terminal. Copper and aluminum foils were used as the current collector of the anode and cathode respectively. Metal foil makers make heat treatment of half annealing in addition to the deoiling of the foil before shipping to the battery companies since any contaminants on the foil surface have a harmful effect on the bonding with the electrode. Bonding between metal foil and the ceramic active materials is notoriously poor, so special care must be taken to control the surface condition of the metal foil.

Lithium ion batteries are produced in the discharged state, so the voltage is not detected when we measure the voltage after assembly process. The batteries need to be activated. We call this activation procedure the formation process, which is the initial charging of batteries. A protective film is formed on the surface of carbon anode by the chemical reaction of carbon with the electrolyte during the formation process, and this film is called the solid electrolyte inter-phase (SEI) film. Physical and mechanical property of this film determines the performance of the lithium ion battery throughout its usage time. The quality of SEI film is related to the materials, cell design and formation process. Prior and subsequent to the formation, aging process is conducted to give the electrolyte enough time to wet the electrode and make the proper electrolyte channels. For the most part, less than 24 hours is sufficient to stabilize the electrolyte channels. However, that's not the case in lithium ion batteries. In actuality, one of the prominent features of the lithium ion battery is its long aging period.

As compared with other secondary batteries, a very thin separator is used to compensate for the low ionic conductivity of the organic solvent electrolyte. Because of the thinness of the separator, it can be easily punctured by any protrusions formed by metallic impurities. To insure the safety of lithium ion batteries, any cells which have the possibility of the separator puncture caused by metallic impurities should be eradicated before shipment. Aging period is used for this purpose along with the original purpose of electrolyte channel stabilization. Aging period is also called the qualification period by virtue of its function of screening out suspicious cells.

Copper, nickel and iron are impurity metals, the amount of which we have to control for the prevention of the short. Copper is likely to come from the slitting process of the anode. The metal can of the cylindrical battery is coated with nickel to prevent the corrosion of steel. Nickel comes from the cylindrical can surface in the beading and crimping process, which is the sealing process by combining the can with a top cap mechanically. Iron mainly comes from the equipment. These metallic impurities are clumped together under electric field and agglomerated by the process of field-affected diffusion. Sony's experimental result showed that it would take one to 3 days for nickel, 7 days for copper and 14 days for iron to complete the field-affected diffusion. Sony multiplied 14 days by the safety factor two to get 28 days aging period. That's how Sony determined 28 days aging period. Aging period has been reduced by the close control of metal impurities.

05

:

Performance improvements

1) Increase in energy capacity

(1) The graphite anode

Since the introduction of lithium ion batteries by Sony in 1991, we had witnessed the steady improvement in the cell performance. Sony's lithium ion battery exhibited cycle life of over 1,000 cycles, which was more than three times higher than its competitor, the nickel metal hydride battery. However, the running time of a lithium ion battery was actually shorter than that of a nickel metal hydride battery because its energy density was lower than that of a nickel metal hydride battery. Sanyo decided to rectify this adverse situation prior to the entry into the lithium ion battery market.

Sanyo developed a lithium ion battery whose energy density is larger than that of a nickel metal hydride battery by changing the anode material from hard carbon to artificial graphite, whose increase in crystallinity caused its energy capacity to increase almost twofold. However, there was one problem associated with the use of high crystalline graphite. Residues of propylene carbonate molecules, ring type electrolyte solvent, block the entrance of graphite, which leads to the rapid decline of cycle life. This phenomenon is called the cointercalation. As a remedy to the cointercalation phenomenon, Sanyo replaced propylene carbonate with ethylene carbonate which has smaller molecular structure than propylene carbonate. In problem solving, solving one problem tends to raise another. The problem associated with the adoption of a new electrolyte solvent was that ethylene carbonate, a ring type organic solvent, is solid at room temperature. Since the melting point changes with the composition, chain type organic solvents such as diethyl carbonate (DEC) or dimethyl carbonate (DMC) was added to ethylene carbonate to lower the melting point and to control other pertinent properties like viscosity.

The microstructure of electrodes is related to the cell performance. The more uniform the electrode microstructure is, the better the cell performance is. Uniform distribution of active materials and electrolyte channels results from the active materials itself among other things, so battery makers endeavored to develop graphite with controlled shape and morphology. Artificial graphite is made from soft carbon by heat treatment called graphitization at over 2000°C, which results in the increase in crystallinity. As regards the microstructure of graphite, it has the layer structure and each layer consists of hexagonal honey-comb structure. Depending on the configuration of the graphite layer, there are two kinds of microstructure: onion skin type and radial shape type. In the carbon fiber used in fiber reinforced composites, each layer is parallel with each other like the onion skin. In contrast, graphite used in the battery anode should have a radial shape microstructure because radial shape structure provides the site of the entry of lithium atoms. The shape of crystal structure is determined by precursors of carbon. When polyacrylonitrile (PAN) based precursor is used in making carbon, the onion skin type graphite is formed after graphitization of carbon. In order to make graphite with radial shape microstructure, mesophase pitch based precursor should be used.

The powder morphology of graphite also plays an important role in the battery performance as it determines the overall structure of the electrode. The electrolyte fills the vacant space of the electrode. The continuous liquid structure formed by the electrolyte is called the electrolyte channel. Uniformity of the electrolyte channels insures the consistent battery performance. In this respect, the morphology of the graphite powder is important, for it determines the microstructure of the electrode. Panasonic developed the spherical graphite called mesophase carbon micro-bead (MCMB), and Toshiba Battery made a graphite fiber called mesophase carbon fiber (MCF).

(2) Cathode materials

Anode materials determine the characteristics of batteries. When the anode material is changed, the whole system of the battery may have to be modified accordingly. In contrast, we can change the cathode materials without significantly influencing the battery system. Moreover, the cathode materials play a critical role in improving the cell performance. Therefore, battery makers tend to put more emphasis on the development of the cathode materials rather than anode materials for the realization of the expeditious improvement of the cell performance. In actuality, most of Japanese battery makers developed and used their own cathode materials in 1990s. However, the internal use of their cathode materials was limited to 40-60%, which reflects their effort not to get left behind in the development information related to the cathode materials. They did not ignore the outside source not to get isolated in the battery materials industry.

• The LCO surface coating

Because of the layer structure of LCO, the surface has the polar nature in that the oxygen layers have the negative charge and cobalt and lithium layers have the positive charge. The electrolyte solvent molecule of the polar nature is attracted to the location of the opposite charge by Coulomb force, in the process of which the electrolyte is disintegrated. This phenomenon is one of reasons for the cycle life reduction. Aluminum is coated on the surface of LCO powder to eliminate the polar characteristic of the cathode surface. Aluminum also acts as the diffusion barrier of the cathode material. Samsung SDI applied aluminum coating on the surface of LCO powders in 1999. As a more advanced surface coating, Japanese battery makers applied titanium (Ti) and zirconium (Zr) coating on the surface of LCO in the early 2000s.

• The nickel base cathode materials

In the late 1990s, portable electronics companies demanded a longer running time. As an example, laptop computer makers wanted their batteries to last for a 5 hour flying time from New York to San Francisco, which is equivalent to 3Ah of 18650 cylindrical batteries. At that time, the energy capacity of 18650 cylindrical batteries was in the range of 1.6-1.8Ah. Therefore, new materials with high specific capacity were required to meet the wish list of portable electronics makers. In response to the market demand, battery makers were beginning to develop high energy capacity cathode materials.

The LNO ($LiNiO_2$) was a good candidate for the next generation cathode material because of the fact that it's energy capacity is higher than that of LCO. However, the problem relative to the LNO was the safety. It becomes dangerous under overcharging condition. If lithium atoms are taken out from the lithium layers of the cathode material excessively under overcharging condition, nickel atoms move to the lithium layers, causing the crystal structure to change from the layer to the cubic structure by spewing out the excessive oxygen atoms. Cathode materials transform from the layer to the cubic structure since the cubic structure is more stable than the meta-stable layer structure thermodynamically. The kinetic barrier that maintains the meta-stable layer structure is destroyed by overcharging.

The tendency of the metal atoms to move to the lithium layers is stronger in LNO than in LCO, which makes LNO more vulnerable to fire or explosion than LCO. This phenomenon can be alleviated by doping such metals as cobalt, manganese and aluminum. These metal atoms in the metal layers can hold the nickel atoms firmly. There are various combination of nickel, cobalt, manganese and aluminum. Among these various nickel base cathode materials, NCA was the first nickel-base cathode material commercialized. The typical composition of NCA is $Li(Ni_{0.8}Co_{0.15}Al_{0.05})O_2$. The main role of cobalt and aluminum is to prevent nickel atoms from leaving the metal layers. The addition of small amount of aluminum decreases the impedance significantly, which makes the NCA battery suitable for the high power application. The nature of low impedance of NCA makes the NCA battery dangerous when the energy capacity is above a certain level. Panasonic provides 18650 cells incorporated with NCA cathode material to Tesla Motors and LG Chemical and Samsung SDI supplies 18650 cells with NCA cathode material to Chinese electric vehicle makers. These NCA cells manage to maintain the integrity and safety because the cell capacity is just over 3Ah.

The NCM cathode material was developed to enhance the safety of NCA batteries. NCM batteries are most commonly used in a wide spectrum of applications ranging from mobile phones to electric vehicles. One example of the composition of NCM cathode materials is $Li(Ni_{0.5}Co_{0.3}Mn_{0.2})O_2$. Development of NCM cathode material is based on the concept that alloying could create the material with the combined advantage of each material. In other words, the design philosophy is to have the combined effect of the high energy of LNO, high rate performance of LCO and high safety of LMO. We may think that blending of these three materials would give us the cathode with a combined optimum property. However, if we blend these materials together, undesirable features of three materials happen to be combined. Therefore, we have to resort to coprecipitation process to combine the desirable features of three materials. In the coprecipitation process, we make the precursors which are in the form of hydroxide, and then mix the precursors with the lithium carbonate to induce the solid-solid reaction by sintering process. After sintering, it becomes as hard as the brick. Crushing and classification procedures are adopted to obtain the NCM powder with specific powder size and distribution.

Unlike LCO which has a simple primary structure, the NCM powder is the agglomerate of small powders, which is called the secondary structure. Because of the rather peculiar microstructure of NCM compared to the simple primary structure of LCO, it's hard to remove the moisture once it is trapped inside the powder. Therefore, more often than not, dry room condition is required in mixing.

If we apply the surface coating to the powder surface of NCM, we can handle NCM powder like LCO without worrying too much about moisture content. Some companies in Korea, with this in mind, coated the powder surface with LFP and carbon black by mechanical alloying method using the Japanese equipment called Nobilta for the improvement of the handling property. In the surface coating of LFP on the surface of NCM by mechanical alloying method, a small amount of titanium oxide (TiO_2) is added, the function of which is to crush the LFP powders like balls in a ball mill. GSEM and KoKam of Korea were in possession of this surface coating technology and this technology was transferred to Dow Chemical in 2009. LG Chemical also has a similar surface coating technology.

2) Price competitiveness

(1) The natural graphite anode

Graphite is more expensive than carbon because graphite is made from soft carbon through high temperature graphitization process. Sometimes, over 3000°C graphitization is needed to acquire high capacity graphite. In contrast, natural graphite is obtained in a graphite mine, which explains why natural graphite is much cheaper than artificial graphite. The key to the use of natural graphite in lithium ion batteries is to refine the natural graphite by eliminating impurities and controlling the morphology and crystal structure so that it fits into the requirements of lithium ion batteries. Sanyo succeeded in developing natural graphite for lithium ion batteries through the finding of appropriate refining process.

Replacement of artificial graphite with natural graphite was not a simple change. Use of natural graphite required the change of the binder and production methodology. Specific surface area of natural graphite is 3-8m^2/g whereas the specific surface area of artificial graphite is 1m^2/g. Since natural graphite has high specific surface area, we need a more powerful binder to acquire a satisfactory binding force. Styrene butadiene rubber (SBR) was adopted as the binder for natural graphite and carboxy methyl cellulose (CMC) was used as the surface active agent. Since water is used as the solvent instead of organic NMP solvent to make a coating slurry, we call the SBR binder a water-based binder. When we use PVDF as the binder, polymer chains of PVDF wrap around active materials. However, the working mechanism of SBR is different from that of PVDF. Graphite powders are practically coated with CMC, and SBR connects graphite powders by way of point contacts. The SBR molecule acts like a spring that connects graphite powders. Because of the difference in the working mechanism, we can achieve the same binding force with a small amount of SBR binder.

Natural graphite complicates the production method as a result of its tendency to generate a large amount of gas. The SBR binder generates five times more gas than the PVDF binder in the formation process. Therefore, gas should be removed before sealing the battery. Precharging process was introduced for this purpose. In precharging process, gas is removed by charging the battery after the injection of approximately 30% of the electrolyte. In addition, high temperature aging process was used to make a sound SEI film on the anode surface.

Korean battery companies got left behind in the introduction of water-based SBR binder. In effect, Chinese battery companies adopted water-based SBR binder ahead of Korean companies. That's because the piping system of the coating equipment didn't allow them to use water as the coating solvent since the pipes were made of low carbon steel instead of expensive corrosion-resistant stainless steel. When Korean battery companies designed their first mass production plant in 1998, they thought that the water-based SBR binder could not replace PVDF binder because of the poor performance of natural graphite. Based on their thoughts on the future lithium ion battery technology, Korean battery companies used inexpensive low carbon steel pipes in the piping of the coater. Low carbon steel is vulnerable to corrosion, so they had to wait until the next battery plant is constructed. In the meantime, Chinese battery companies perfected their skills in handling water-based SBR binder in 2000s. They extended their knowledge and experience in the application of SBR binder from the graphite anode to the area of LFP cathode. Chinese battery companies managed to adopt water-based SBR binder in both the anode and cathode. As a result, Chinese battery companies was able to eliminate fire-risk NMP solvent from their battery plants.

(2) The electrolytic copper foil

High price of lithium ion batteries had been a continuing stumbling block to market expansion throughout the 1990s. For the reduction of cost, it was required to reduce the materials cost simultaneously with the increase in the production yield. With regard to the production yield of Japanese lithium ion battery makers, yield rate of 30% was maintained for quite a long time and then jumped to 90%. In contrast, production yield of Korean companies exhibited a steady growth from the initial yield of 70% thanks to the merit of followers.

Battery materials are classified into active materials and inert materials. We call the anode and cathode materials as the active materials because they determine the cell performance. On the other hand, metal foils and the separator are called inert materials because they don't have any direct influence on the cell performance. Reduction of the inert materials volume provides us with more space for active materials. Therefore, if we diminish the amount of inert materials, we can increase the energy capacity proportionately. Moreover, if we can use cheaper inert materials without any sacrifice of cell performance and safety, we can reduce the cost. Generally speaking, the activity that reduces the cost should start with inert materials to minimize the side effects accruing from the change of active materials.

Japanese battery makers found a way to reduce the cost of copper foil. There are two kinds of copper foil depending on the manufacturing method: rolled and electrolytic copper foil. Rolled copper foil is made by such typical metallurgical manufacturing process as melting, ingot casting, hot rolling and cold rolling whereas electrolytic copper foil is made by a simple electro-winning process. With regard to the raw materials of copper foil, the expensive electrolytic copper cathode should be used to maintain the foil quality in rolled copper foil. In contrast, cheap copper scraps are commonly used in making electrolytic copper foil. Hence, rolled copper foil is much more expensive than electrolytic copper foil intrinsically.

In spite of the obvious cost advantage of electrolytic copper foil, Sony had to use expensive rolled copper foil because electrolytic copper foil was not compatible with the requirement of lithium ion batteries at that time. Electrolytic copper foil had different surface roughness profile on both sides, and the mechanical property of the foil was not isotropic due to the elongated grain shape. Toshiba Battery came forward to make electrolytic copper foil for lithium ion batteries in collaboration with the major electrolytic copper foil maker in Japan. They developed electrolytic copper foil exclusively for lithium ion battery which has the identical surface roughness profile on both sides and isotropic mechanical property by the formation of round shape grains. Once the cost-effective electrolytic copper foil was available in the market, it didn't take long to replace the expensive rolled copper foil.

Another advantage of electrolytic copper foil is that it's rather easy to make thin copper foil. In rolled copper foil, a more sophisticated rolling machine is required when the foil becomes thinner, and the difficulty of making foil increases as the thickness of the foil is reduced. However, that's not the case in electrolytic copper foil. The trend of the portable electronics market is to reduce the thickness of metal foil to increase the energy density. Copper foil makers responded to the market demand by improving the technology of making thin electrolytic copper foil.

3) The battery types

(1) Cylindrical batteries

The cylindrical battery is the most common type of batteries regardless of the battery chemistry. That's why the size was standardized internationally, but lithium ion battery does not comply with the international standard. In fact, there had been comprehensive discussion on the standardization of lithium ion batteries, but standardization was not implemented because of the Sony's objection. Nonetheless, the size of 18650 had been used in a wide spectrum of applications since Sony's use in the camcorder in 1992. The cylindrical battery exhibits higher energy capacity than any other types of batteries by virtue of its favorable jelly roll configuration. Since it is amenable to automation, the cylindrical battery has the highest production rate. One of the key features of cylindrical lithium ion batteries is the improvement of the safety by the introduction of the CID. The CID is a sophisticated vent which prevents thermal runaway under overcharging condition by cutting off the aluminum tab of the cathode, but it increases the impedance of cylindrical batteries by interrupting the electron passage.

In the beginning, the energy capacity of 18650 cylindrical batteries was 1.2Ah. Change of the anode material from hard carbon to graphite increased the energy capacity to 1.5Ah. In the 1990s, energy capacity increased by the interval of 0.2Ah through the increase of electrode density and the reduction of the thickness of the separator and aluminum and copper foil without any specific change of the electrode materials. Increase in the energy capacity to over 2Ah makes the jelly roll of the battery vulnerable to collapse under the abused condition. A center pin, which acts as the supporter of the jelly roll, was introduced to prevent the collapse of the jelly roll. The shape and design of the center pin were determined through close examination of the degradation process of the jelly roll. At first, a steel or ceramic rod was used to support the weight of the jelly roll. However, the safety test showed that in some cases a center pin aggravated the safety of the battery. Subsequent investigations revealed that the center pin should act as the channel of the smoke like the chimney in addition to the supporter of the jelly roll. It was found that use of a ceramic or stainless steel rod aggravated the situation by blocking the smoke pathway in the abused condition, so the rod was replaced by the tube. Another finding revealed that the safety level of lithium ion batteries changed depending on the shape of the cross section of the tube. When the jelly roll deteriorates, the cross section of the center of the jelly roll takes the shape of a heart. Based on this observation, metal tube whose cross section has the heart shape was used as the center pin. In sum, the metal tube center pin with a heart shaped cross section acts as both the supporter of the weight of the jelly roll and a chimney through which smoke formed in the battery under the abused condition is to be discharged outside expeditiously.

As a way of hermetic sealing of the battery, beading and crimping method is used, in which beading is the process to make the neck on the can at the site where the top part of the jelly roll is placed and crimping is the process that combines the can with a top cap of the battery. Unless the load applied to the can and a top cap during beading process is uniform, it can cause stress corrosion during its usage. Stress corrosion refers to cracking caused by the simultaneous presence of tensile stress and a specific corrosive medium. Beading and crimping process is one of the key technologies associated with cylindrical battery manufacturing. It takes several years to perfect this technology, so beading and crimping process is considered as one of the entry barriers of cylindrical batteries. Panasonic, Sony, Samsung SDI and LG Chemical belong to the cylindrical battery makers.

Heat dissipation problem arises when the number of turns of the jelly roll increases due to the increase in energy capacity and the concomitant increase in the size of the battery. In case the electrode thickness is not uniform, protrusions on the electrode act as the welding points, at which electrodes and the separator are welded together locally and burnt in high current condition. In the high voltage applications such as electric vehicles and energy storage systems, an arc can be generated at a CID by the concentration of the total pack voltage on the CID of a single activated cell. An arc at the open CID can easily lead to ignition of hot electrolytes. Ford in collaboration with Samsung SDI of Korea found that arcing actually happens at the site of CID opening and the minimum voltage at which arcing can be formed is around 220V.

Tesla Motors has been using 18650 cylindrical batteries supplied by Panasonic of Japan in the electric vehicles. Tesla Motors' choice of Panasonic's 18650 cylindrical batteries is understandable in that these cells have the highest energy density and the lowest price among various types of lithium ion batteries. Moreover, Panasonic is the leading company in 18650 cells in terms of the uniformity of the cell capacity, which is the most important criterion in the electric vehicle batteries. Tesla Motors, which was founded in 2003, introduced its first electric vehicles in 2008. The first model called Roadster employed Panasonic's 18650 cells incorporated with the LCO cathode. The LCO cathode was replaced by NCA in the subsequent vehicle models to increase the driving range. Tesla's vehicle adopting a 310V motor used 5,544 cells. In this system, fourteen modules are connected in series. Each module contains 396 cells, among which 66 cells are connected in parallel to form a sub-module, and six sub-modules are connected in series to make a 22V module. In a 360V system, a battery pack contains 7,104 cells. Sixteen modules are connected in series, and each module contains 444 cells, among which 74 cells are connected in parallel to form a sub-module, and six sub-modules are connected in series to make a 22V module. As the voltage of a module is well below the voltage of an arc generation in CID, Tesla Motor's battery is immune from the danger associated with CID opening. Confident that 18650 cells are not in danger of generating an arc when the CID is opened, Panasonic adopted the double CID system to reinforce the safety feature. The CIDs are installed at the bottom as well as the top part of the cell. Confidence accumulated through the track record of the use of high power 18650 cells in an electric vehicle environment enabled Tesla Motors as well as Panasonic to construct the Gigafactory in Nevada. Tesla Motors is keen to increase the size of a cylindrical cell from 18650 cylindrical batteries to 20700 (diameter 20mm, height 70mm) cylindrical cells not only to increase the energy capacity from 3.2Ah to 5V but also to decrease the number of cells in a pack. Since individual cells should be controlled in lithium ion batteries, decrease of number of cells reduces the pack cost. Tesla Motors may switch from small cylindrical batteries to a large prismatic battery after 2020 to increase the efficiency of the pack design further. Panasonic has the best technology in a large prismatic battery.

(2) Prismatic batteries

Up until mid 1990s, the operating voltage of mobile phones was 5V. Five nickel metal hydride prismatic cells were connected in series to make a 6V power pack while two cylindrical lithium ion cells were connected in series to make a 7.2V power pack. Nickel metal hydride batteries had better voltage compatibility than lithium ion batteries. In these circumstances, Motorola introduced 3V mobile phones to the market by developing a 3V power amplifier, which was the last 5V component in the mobile phone. The 3V mobile phones eliminated the need for the serial connection of cells in lithium ion batteries. Mobile phone makers preferred a 1 cell pack of lithium ion batteries to a 3 cell pack of nickel metal hydride batteries because of the compactness of the pack and a longer running time. By the change of operating voltage of mobile phones, prismatic lithium ion batteries emerged as the major battery in the mobile phone market.

In the 1990s, the small prismatic nickel metal hydride battery was in use in portable electronics, which was commonly called a gum battery because of the size and shape similarity to the chewing gum. The trend of the preference of compactness was extended to lithium ion batteries. Battery makers applied prismatic lithium ion batteries to the mobile phone.

The first model of prismatic batteries had the dimension of 30mm width, 48mm height and 8mm thickness, which we referred to as a 30×48 foot print. Increase in the energy density enabled us to decrease the thickness to 6mm. As the competition for thinner mobile phones intensified, thickness was reduced to down to 3mm with the concurrent improvement of the metal can production technology. Another trend for mobile phones in the late 1990s and early 2000s was wider and more functional phones. In order to meet this demand, the 34×50 foot print was introduced since more energy could be contained in this size.

The role of the metal can in addition to its conventional role of the container of jelly roll and the electrolyte is to apply the load to the electrodes, without which the battery would not function properly. It is like squeezing the toothpaste out. The load applied to the electrode should be uniform for the consistent cell performance. Uniformity of the load depends upon the foot print, among which the foot print of 30×48 and 34×50 was proven to spread the load with great uniformity. Pouch batteries have the advantage of design flexibility over prismatic batteries, which is often misinterpreted as the amenability of the pouch batteries to designing any shape of batteries. Of course, it is not quite the case. Truth be told, the design flexibility is limited to the flexibility in a foot print. Electrodes are laminated with the separator in pouch batteries, so pouch batteries don't have to rely on the pressure from the can for the proper application of the load. The reason for the use of a thicker can in prismatic batteries compared to the cylindrical batteries is due to the fact that a cylindrical can is more effective in applying the load geometrically in addition to the strength difference between steel and aluminum. As the size of prismatic batteries was diversified, the importance of the proper foot print has been faded. However, it should be noted that the optimum foot print of the prismatic batteries not only insures the reliability but maintains the soundness of the battery by restraining the jelly-roll deformation.

The function of the prismatic battery vent, the role of which is to release the gas in case of overpressure, is simpler than that of CID of cylindrical batteries since it is devoid of the current interrupting function. It is also possible to install the CID in large prismatic batteries to secure the safety, but the thickness should be over 15mm for the CID to be implemented in the prismatic batteries. Whereas most of the battery companies use the clad vent installed on the top of prismatic batteries, some battery companies adopted the side vent, which is a L shaped notch indented on the top part of the side of the aluminum can. Difference between the clad and side vent is not limited to the configuration and the location. The working mechanism of the side vent is different from that of clad vent. Whereas the built-up gas pressure inside the can tears off the vent in a clad vent, mechanical force generated in the distortion of a jelly roll is the major force to break the notch in the side vent, so the accurate control of the breakage force of the vent becomes more difficult in the side vent. One of the ways to increase the energy density is to reduce the free space inside the metal can. However, if the free space inside the can is too small, interference of battery parts is prone to take place in the prismatic battery with a side vent, which can lead to field incidents. In effect, the field incident of prismatic batteries with a side vent was a serious problem in 2004-2005.

Panasonic, PEVE (currrently Primearth), Samsung SDI and GS Yuasa develop prismatic batteries for the hybrid electric vehicle (HEV) and electric vehicle (EV) application. As regards the cell performance and the uniformity of cell capacity, Panasonic and PEVE seem to be leading the prismatic battery industry. GS Yuasa made a joint venture called Lithium Energy Japan (LEJ) with Mitsubishi Motors in 2007 and made another joint company called Blue Energy with Honda in 2009. Prismatic battery technology of GS Yuasa through Blue Energy and LEJ show that GS Yuasa has strength in the tool and part design. As for Samsung SDI, its strength comes from the experience to work with major automobile makers.

High performance EV batteries can be reused in other applications after the elapse of over ten years life because these batteries still hold over 80% of the initial energy capacity. These batteries can be reused in such application as energy storage system (ESS). In other words, EV batteries are designed to be reused in other applications after termination of their lives in the EVs. The prismatic battery is the best choice in terms of the reusability because sealing by laser welding makes it possible to last for over twenty years. German automobile makers along with Ford prefer the prismatic battery to the pouch battery not only because of its rigidity but also because of its amenability to standardization. Automobile makers know from experience that the most effective way to reduce the cost is the standardization.

In order to circumvent the difficulty associated with flat winding of prismatic batteries, Chinese battery makers developed hybrid prismatic batteries in which a jelly-roll of the pouch battery made by stacking and folding process was used. The top cap and the aluminum can were sealed by laser welding. Hybrid prismatic batteries have been used in the Chinese electric vehicles, and non-Chinese makers are beginning to take interest in the hybrid prismatic batteries. For example, Samsung SDI developed the hybrid prismatic batteries for the battery samples of Apple's self-driving cars in 2016.

(3) Pouch batteries

The first company which tried to commercialize pouch batteries was Valence Technology of the USA. In the early 1990s, a LPB was once expected to be a dream battery which could surpass lithium ion batteries in terms of cell performance and cost. Expectation was so high that many portable electronics companies endeavored to have access to that technology in advance by funding the development of the LPB.

The LPB uses lithium metal foil as an anode, and the electrolyte and a separator are replaced by a polymer electrolyte film. The basic idea of a LPB is to improve the safety of lithium secondary batteries by eliminating the interaction of lithium metal with liquid electrolytes. However, glass transition temperature of a polymer electrolyte was so high that it was not operable at room temperature. Plasticizers were added to improve the low temperature performance, but addition of plasticizers to the polymer electrolyte not only aggravated the mechanical property of the polymer electrolyte but increased the interfacial resistance between plasticizers and the polymer matrix. Although a polymer electrolyte improved the safety of lithium metal, it couldn't provide the performances that could compete with lithium ion batteries. In order to resolve these problems, lithium metal was replaced by graphite anode, which led to the lithium ion polymer battery (LiPB). However, low ionic conductivity of a polymer electrolyte decreased rate performance and low temperature discharge capability.

Bellcore of the USA developed plastic lithium ion (PLI) batteries by replacing a polymer electrolyte by a polymer film impregnated with liquid electrolytes. PLI batteries improved rate and temperature performance at the expense of energy density. However, low energy density of PLI batteries in comparison with lithium ion batteries made it difficult for the PLI batteries to compete with prismatic lithium ion batteries in the mobile phone market where slim design of the battery becomes the primary importance.

It was Sony that came up with the idea on how to compete with prismatic lithium ion batteries in the mobile phone market. Sony introduced a thin separator to have better performance than prismatic lithium ion batteries. The only conspicuous difference between Sony's batteries and prismatic lithium ion batteries was the casing. That's why it was called the pouch battery. The hurdle to the commercialization of PLI batteries was bulky size caused by low energy density, and because of this intrinsic limitation, PLI batteries had to experience difficulty in penetrating into a mobile phone market. Sony's introduction of a thin separator in conjunction with gel type electrolytes instead of a bulky polymer film impregnated with liquid electrolytes enabled pouch batteries to enter the mobile phone market for the obvious advantage of compactness. In actuality, the neck-and-neck competition between Sanyo's thin prismatic battery and Sony's pouch battery in the area of mobile phone was intense and intriguing. What matters most is that their competition made a great contribution to the shaping of the battery industry.

LG Chemical's strategy was to develop pouch batteries that could increase the rigidity of the battery. In essence, the target was to develop the hybrid battery which had the combined feature of the metal can prismatic battery and the pouch battery. Pouch batteries have high energy density, high rate capability as well as good heat dissipation capability. However, pouch batteries have the intrinsic weakness of lack of robustness. In order to compensate for the weakness of the conventional pouch batteries, LG Chemical reinforced the structure of electrodes through their proprietary electrodes assembly technology.

Introduction of smart phones by Apple in 2007 changed the competitive composition of the battery. Smart phones require over 2-3Ah energy capacity with a large foot print, which enabled pouch batteries to have a competitive edge over prismatic batteries. The advantage of the design flexibility of the foot print in pouch batteries becomes more pronounced in the embedded battery.

In the battery assembly technology, the most difficult process of the cylindrical battery is beading and crimping process. Poor beading and crimping technology would lead to electrolyte leakage and short cycle life. In case of prismatic batteries, flat winding of electrodes and laser welding technology are the most critical process. Without good winding and laser welding technology, cell performance would be degraded prematurely. Therefore, it will take at least 3-5 years to acquire the reliable battery manufacturing technology in cylindrical and prismatic batteries.

With regard to the pouch battery, pouch battery was created with the emphasis placed on eliminating any critical process, so companies without any tangible experience in the battery business can make pouch batteries. The pouch battery is defined as the battery without any parts and a can, which implies that the battery performance is almost identical to that of electrodes. In cylindrical and prismatic batteries, parts and a can lower the cell performance. That's why pouch batteries exhibit better cell performance than the prismatic batteries and, more importantly, cell performance of the pouch batteries levelled off regardless of the battery makers. As regards the standardization, the beauty of the pouch battery is not to standardize.

Ease of manufacturing of the pouch batteries implies that the entry barrier of the pouch battery is very low. Since it is the material not the manufacturing technology that makes the difference in the cell performance, the leading pouch battery maker should possess the superior materials technology by which to make a high entry barrier, especially the electrolyte technology. The superior cell performance of pouch batteries should be utilized to lower the cell price. The pouch battery makers should be able to make the battery whose cell performance is tantamount to that of prismatic battery even when they use low quality materials. In other words, one of the strategies pouch battery makers should take is to acquire the cost competitiveness by adopting cheap materials. Therefore, when we evaluate the pouch battery makers, the cost structure should be the primary criterion.

(4) Large format batteries

Change from small batteries of mobile phones and laptop computers to large format batteries used in electric vehicles is not just the simple extension of the size. The design of large format batteries encompasses the increase of power and cycle life without expense of the safety. The design change of electrodes and the tab to facilitate the ion and electron movement is implemented by enlarging the tab area and reducing the electrode thickness concurrently with the increase of electron conducting materials. Thickness of metal foil and a separator should be maintained at a certain level to supply the electrons profusely and to hold enough electrolyte, without which the battery will not be able to last for over ten years.

Requirement of the large tab area in a high power application changes internal configuration of cells. In the prismatic batteries, a jelly roll is inserted sideways to combine with terminals on both ends of the jelly roll, resulting in a floating configuration of a jelly roll. One of the most convenient ways to increase the energy capacity is to increase the electrode density by roll pressing process, but this method should be applied to a large format battery with careful consideration because battery life is inversely proportional to the electrode density. Low electrode density enables the battery to cope with the electrode stress smoothly and hold enough electrolytes to maintain the stable electrolyte channels in the electrodes. Electrode density also has associated with heat dissipation capability, which is a critical parameter in the design of large format batteries. It is commonly known that prismatic batteries have a better heat dissipation capability than cylindrical batteries. However, when we compare the heat dissipation capability between cylindrical and prismatic batteries, the result shows that it depends on the thickness of the prismatic batteries. According to the testing result of Toyota, prismatic batteries exhibit better heat dissipation capability than cylindrical batteries only when the thickness of the prismatic battery is below 15mm.

One of the limitations of lithium ion batteries is the size of the battery that can be constructed with the reasonable confidence on the safety. The current status of the maximum energy capacity of the cell is below 100Ah. However, such Chinese battery companies as Calb, Winston, GBS and Sinopoly had taken a different approach to increase the energy capacity of the cell. Combination of a plastic can with the high safety lithium iron phosphate (LFP) cathode materials enabled them to make large format batteries of over 100Ah energy capacity. Chinese battery makers replaced the aluminum can with the plastic can in the hybrid prismatic battery to hold more electrodes.

Demonstration of large energy capacity batteries in a real environment such as a golf cart was rather successful, but this battery suffered from intrinsic limitations. Owing to the poor heat dissipation capability of the plastic can, temperature of the cell rises rapidly with the increase of the current rate. Temperature of the cell reaches over 40°C at 2C discharge, and to make matters worse, temperature measurement of the cell becomes difficult because of the use of the plastic can instead of the metal can. Temperature is important in SOC calculation, so the plastic can prismatic battery cannot be used in a sophisticated application.

As compared with other lithium ion batteries, the plastic can prismatic battery is manufactured in a rather rough environment. In the assembly process, dry room or dry box is not used with the exception of the electrolyte injection process. Aging period is very short since aging period is not utilized as the qualification period, which means that defective cells are not screened out during aging. In spite of the intrinsic roughness in manufacturing, plastic can prismatic batteries keep replacing lead acid batteries in the golf cart application because of the cost competitiveness and the large energy capacity.

4) High power batteries

(1) The battery market trend

When Sony commercialized lithium ion batteries in 1991, the movement of Sanyo and Panasonic was important since they had the capability to change the market. Panasonic was the inventor of nickel metal hydride batteries along with Toshiba Battery. The nickel metal hydride battery was as new as the lithium ion battery at that time. It was commercialized in 1990, just one year before the market entry time of lithium ion batteries.

The senior managers of Panasonic wanted to focus on nickel metal hydride batteries as they wanted to return their investment on nickel metal hydride batteries in a short period of time. However, they could not ignore the potentials of lithium ion batteries, so they decided to go along with the market trend of lithium ion batteries with the emphasis placed upon nickel metal hydride batteries.

As for the mobile phone companies, Motorola, Samsung Electronics and LG Electronics led the adoption of lithium ion batteries in their mobile phones. The common denominator of these mobile phone makers was that they employed code division multiple access (CDMA) system. CDMA was more compatible with lithium ion batteries than global system for mobile communications (GSM) in that CDMA needed more energy and required less high pulse power. The operating voltage of the mobile phone at that time was 5V, so serial connection of five 1.2V nickel metal hydride batteries gave a 6V power pack whereas two serial connection of 3.6V lithium ion batteries made a 7.2V power pack. Voltage difference of lithium ion batteries was one of the problems facing lithium ion batteries because the efficiency of DC/DC converter was not satisfactory at that time. Furthermore, prismatic nickel metal hydride batteries had less dead space than two 18650 (diameter 18mm, height 65mm) cylindrical lithium ion batteries. Comparing lithium ion batteries with nickel metal hydride batteries in terms of the pack space, nickel metal hydride batteries took up less space and exhibited a better configuration.

Future of lithium ion batteries seemed bleak unless some measures were taken to make mobile phones more compatible with lithium ion batteries. Motorola, a leading CDMA mobile phone company, took actions. Motorola devoted its efforts to develop 3V mobile phones among other things. As a result of Motorola's efforts, a one cell pack of the lithium ion battery became a reality. Pursuant to the operating voltage change from 5V to 3V, the batteries of mobile phones were beginning to change from prismatic nickel metal hydride batteries to prismatic lithium ion batteries, especially in CDMA phones.

Motorola's development of a 3V mobile phone will justifiably be recorded as one of the most important events in the secondary battery history since it enabled lithium ion batteries to make an inroad into nickel metal hydride battery market. Unlike Qualcomm's CDMA, GSM required high pulse power. Furthermore, GSM phones were cheaper than CDMA phones. Therefore, nickel metal hydride batteries with the advantage of cost and high pulse power capability dominated GSM mobile phone market even after the appearance of 3V mobile phones. Nokia was the leading company in GSM mobile phones, and used Sanyo's prismatic nickel metal hydride batteries exclusively through the partnership with Sanyo. They collaborated with each other from the product planning stage. In so doing, Sanyo was able to develop batteries reflecting the customer's needs one step ahead of others and sold its batteries at a premium price. That was the strategy of Sanyo, the top battery maker in the 1990s. The partnership between Nokia and Sanyo enabled both companies to be ahead of the curve.

Based on their confidence on the safety of lithium ion batteries and the accumulated track record of safe usage of lithium ion batteries for almost ten years, battery makers improved high pulse power capability for the application in GSM mobile phones. Pursuant to this improvement, Nokia, the top GSM mobile phone maker, finally decided to switch from Sanyo's nickel metal hydride to lithium ion batteries, and started to use lithium ion batteries other than Sanyo's. In accordance with Nokia's changed policy on the battery suppliers, Panasonic entered a Nokia market, followed by Samsung SDI. Strong bond between Nokia and Sanyo became weaker. Nokia's decision to use lithium ion batteries in

GSM mobile phone was a final blow to nickel metal hydride batteries in portable electronics market. Finally, lithium ion batteries became the battery of portable electronics.

The market size of portable electronics was rather large. Nonetheless there was no room for both of them. Nickel metal hydride batteries left the market for good. Battery history shows that it's very unlikely that batteries with different battery chemistry coexist in the same market. Once the dominance of one battery system manifested itself, dominance prevailed in a short period of time. In general, it is the leading company that adopts the new battery in its portable electronics or electric vehicles ahead of others. Its decision to employ a new power system becomes a sort of standard regardless of the pros and cons associated with the new battery. If the leading company opts for a new battery, most of other companies followed suit not to be left behind in the industry trend.

(2) Power tool batteries

The major factor to limit the application areas of the lithium ion battery was its inability to discharge at high current rate. The safety concern at high current was reflected in the battery design: current was cut off at the current of over 2C rate discharge. Therefore, lithium ion batteries were used in the applications in which batteries don't have to operate the electric motors such as cellular phones, laptop computers and the camcorder, which were called the 3C market.

As time goes by, time has come for the lithium ion battery industry to confront the issue of high power capability, without which expansion of lithium ion battery market would be severely hampered. Confidence of battery makers on the safety of lithium ion batteries through the track record of almost ten years was strong enough to overcome their concern on the safety in high current discharge applications. Finally, battery makers succeeded in developing lithium ion batteries with high power capability in 1999.

Current is the electron flow, so good analogy of the current is the water flow in the water pipe. We need a large diameter pipe to flow a large amount of water in a short period of time. Similarly, the tab area of the electrode should be enlarged as the current increases. Because of the configuration change of the electrode, the coating method was changed from intermittent to strip coating. In the beginning, many tabs were attached to the uncoated area of the current collector to increase the tab area, but this kind of design made it difficult to insure the consistent battery quality. So the total uncoated area was used as the tab.

Electrodes became thinner to reduce the distance of the lithium ion movement, and the amount of electron conducting materials like carbon black was increased. The first target market of high power lithium ion batteries was the power tool in which nickel cadmium batteries had been in use for a long time. In 2003, lithium ion batteries successfully entered a power tool market in competition with nickel cadmium batteries, which was made possible by the improved cost competitiveness as well as high power capability.

Unlike 3C applications like cellular phones, laptop computers and the camcorder, a power tool is the application in which the battery is exposed to severe vibration, so the robustness in the cell design, especially in the welding area, had to be reinforced to withstand the vibration of a power tool. In addition to the difference in the cell design described above, Sony, Sanyo and Panasonic of Japan introduced new cathode materials, NCA and NCM, in their power tool batteries. Sony used a composite of tin and cobalt alloy with graphite in the anode, which was called Nexelion. Japanese battery makers wanted to test new materials in the power tool application. In effect, the power tool was the first application to employ high capacity nickel base cathode materials. Sanyo also improved the safety further by using lithium iron phosphate (LFP) cathode in another version of power tool lithium ion batteries.

A power tool requires discharge power as high as 20C rate, so successful application of lithium ion batteries to the highly dangerous operating condition in terms of the discharge current rate positioned lithium ion batteries to penetrate into more challenging market like hybrid electric vehicles.

(3) xEV batteries

Smog is the compound word of smoke and fog. Unlike London smog, Los Angeles smog is caused by automobile emission gas. Considering this historical background, movement of California to restrict the emission of vehicles is understandable. California mandate stipulates that top seven automobile makers sell zero emission vehicles (ZEVs) if they are to sell their cars in California. The ZEV is a vehicle with no internal combustion engine such as electric vehicles and fuel cell vehicles.

With reference to California mandate, automobile makers protested that California mandate was too unrealistic to comply on account of the extremely high cost of the power source. In response to automobile makers' protest, the state government officials of California announced a concession plan of a partial credit system. They decided to give a partial credit to hybrid electric vehicles: 0.1 for engine dominant HEVs and 0.2 for battery dominant HEVs. An engine dominant HEV is a parallel type HEV in which both the engine and an electric motor power the automobile simultaneously. A battery dominant HEV is a serial type HEV in which an automobile is driven by the electric motor and batteries are charged by an engine. Energy capacity of a battery dominant HEV is larger than that of an engine dominant HEV, which explains the difference in the credit point.

Some automobile makers wanted their hybrid vehicles to have the capability to compete with conventional vehicles in the future market regardless of California mandate, so they endeavored to find a way to design HEVs that could make profits. To design profitable hybrid vehicles, it was necessary to reduce the energy capacity of the power source. Their efforts to reduce the energy capacity of a power source gave rise to power assist HEVs. In a power assist HEV, engine is rotated at a constant rpm, which reduces the gas emission and increases a gas mileage. As more power is needed, the electric motor assists the engine power. When we apply the brake, this power is used to charge the batteries, which is an energy recovery system known as the regenerative braking. Power assist HEVs are specified based on the power, so the energy capacity and the voltage required to operate HEVs change with the battery chemistry. For example, batteries of 7Ah capacity are used for a nickel metal hydride battery while batteries of 4.5-5Ah capacity are sufficient in a lithium ion battery. Power assist HEVs are not compatible with a manual transmission, which made power assist HEVs unpopular in Europe where people enjoy driving a car with a manual transmission. In a driving environment of North America where highway driving is the major driving pattern, a gas mileage drops because of the reduction in efficiency. To sum up, power assist HEVs are efficient in such places as Japan and Korea where traffic jam is frequently encountered.

Major power assist hybrid electric vehicle makers were Toyota, Honda and Ford. They had power train technology of HEVs. After their development of power assist HEVs, they searched for batteries that fit into the requirement of a HEV. Fortunately, D size (diameter 34.2mm, height 61.5mm) nickel metal hydride batteries of Panasonic and Sanyo fit the bill. Toyota formed an alliance with Panasonic, giving birth to Panasonic

Electric Vehicle Energy (PEVE), and Sanyo supplied high power nickel metal hydride batteries to Honda and Ford. PEVE was founded in 1996 and has been producing NiMH batteries for a HEV application. It changed its name from PEVE to Primearth EV Energy Co. to facilitate the merging of Sanyo in 2010. After Panasonic's acquisition of Sanyo, PEVE started to produce prismatic lithium ion batteries for xEVs applications in addition to nickel metal hydride batteries in 2011.

06
:
Battery manufacturing process

1) The plant structure

In general, the lithium ion battery plant is comprised of two separate buildings: a battery production building and a formation & aging building. These two adjacent buildings are connected by a sky-corridor, and cells are moved to a formation building by an automated guided vehicle (AGV). It is a common practice to separate the electrode making and assembly process because the working environment and the nature of process are different. Vertical separation is conventionally adopted in Japan and Korea where the cost of land is high whereas horizontal separation is commonly preferred in North America with a spacious land. As lithium ion batteries are expanded into the field of xEVs, the structure of the plant is diversified.

The formation building is separated from the main production building because the formation room is vulnerable to the outbreak of fire in the formation and aging process. In fact, there occurred incidents of fire in the formation building during the period of the first five years after the inception of the operation of the mass production plant in Japan. From the perspective of the manufacturing process, an electrode making is a chemical process and an assembly is a mechanical process and the formation is an electrical process. In this regard, we can justifiably call the battery manufacturing a comprehensive technology. In the configuration of the battery components inside the battery, the anode and cathode are separated completely by the separator to prevent the short. Similarly, the anode and cathode making facilities are separated completely to prevent cross contamination, which requires two identical sets of equipment respectively. In effect, one of the criteria that differentiate the mass production plant from the pilot plant is whether or not we have the separate production facility of the anode and cathode.

The electrode making process, which is conducted in the first floor of the battery production building, encompasses mixing, coating, roll pressing and slitting. Cells are assembled in the dry room of the second floor, the relative humidity of which is maintained below -40°C. Vacuum drying process, in which the material flow is from the first to the second floor connects the electrode making to the assembly room. The battery assembly process consists of jelly roll making, jelly roll insertion, tab welding, the electrolyte injection and hermetic sealing of the metal can. There are some differences in the assembly process between cylindrical and prismatic batteries. In cylindrical batteries, the electrolyte injection is followed by beading and crimping process. On the other hand, in prismatic batteries we weld the top cap to the aluminum can by laser welding and then inject the electrolyte through an electrolyte injection hole, which is sealed with a small ball after completion of the filling of the electrolyte. After the assembly of cells, we wash the electrolyte smeared on the surface of the can by washing process, without which battery would be vulnerable to corrosion during usage. After washing, cells are moved to a formation room by an AGV.

The formation and aging process consists of formation, aging and screening out defective cells. The formation process, which is the initial charging of the battery, forms a solid electrolyte inter-phase (SEI) film on the anode surface by the chemical reaction of the anode with the electrolyte. The aging conducted subsequent to the formation process stabilizes the electrolyte channels formed on the electrodes. Defective cells and any suspicious cells that may cause trouble in the field are screened out in the aging process. That's why the aging period is also called the qualification period. The formation room was designed based on the automatic warehouse, so it is fully automated and data management system is implemented for battery management such as back-tracking of cells.

2) Electrode manufacturing process

(1) Mixing process

A planetary mill, which is compatible with both anode and cathode materials, has been used in making coating slurry. If we use the ball milling process, which applies high pressure on the materials, graphite particles crumble by the destruction of the crystal structure, so graphite loses its capability to intercalate lithium atoms. In contrast, cathode materials are so strong and rigid that they can withstand any kind of milling methods. Hence, the choice of a planetary mill is based on the proper milling method of anode materials.

In general, a solution mixing method is used in which we make binder solution by mixing polyvinylidene fluoride (PVDF) binder with N-methyl-2-pyrroliodone (NMP) solvent, and then mix the binder solution with powder to make coating slurry. In water based binder using styrene butadiene rubber (SBR) and carboxy methyl cellulose (CMC), mixing is more difficult than PVDF binder on account of the high specific surface area of the powder. It can be alleviated by the powder mixing method in which powder mix is made at first and then mix the powder mix with water to make uniform slurry. We control the slurry viscosity and solid content for the subsequent coating process.

Mixing is the critical process that determines the quality and safety of the cells, so close control of the mixing process is necessitated to insure the quality of the cells. The following example will give you a good idea of how important it is to manage the mixing process with utmost scrutiny. An operator of a mixing process had dropped his personal belongings to the inside of a mixer during pouring of raw materials into the mixer by accident. However, he didn't bother to report to his supervisor what had happened because he thought it wouldn't hurt anything as the object was so small. He thought wrong. This object changed the clearance between mixing blade and the inner wall of a mixer every time a mixer was run. Accumulated foreign objects caused by the repeated drop of small things to the mixer aggravated the situation. Accurately controlled clearance became smaller by the distortion of the shaft and finally the blade and the inside wall of the mixer touched each other. The blade scraped off the surface layer of the mixer wall, and left the scratch on the mixer wall, in the process of which metal powder was formed. As time goes by, scratch became deeper and the amount of metal powder increased, and the metal powder was impregnated into the mixing slurry. The amount of metal powder included in the slurry increased with time. Nevertheless, a mixer was operated for a couple of months unnoticed. Two months after this unreported incident, an operator in the formation room found some abnormal electric behavior by accident. After intensive investigation of the process and equipment, they finally found out that impurities came from a mixer, and to their surprise, found all sorts of foreign objects inside a mixer.

What took them so long to figure out what had happened? Since a formation room is the control center, all the information associated with manufacturing are supposed to be gathered in the formation room. Nonetheless, their failure to detect the continuous inclusion of metallic impurities is due to the fact that stainless steel behaves like an inert material electrochemically. Stainless steel is an alloy steel in which chromium and nickel are added to iron, forming chromium oxide on the surface. This chromium oxide on the surface of the stainless steel hinders the steel from being corroded. The well-designed formation system to sort out cells including metallic impurities becomes ineffective in non-metallic impurities and electrochemically inert materials like stainless steel powder. To compensate for the limitation to detect accidental inclusion of foreign inert materials, sound of machine should be utilized to guard against the inclusion of non-metallic impurities and foreign objects. Operators in the mixing process should have discerned if the mixer is operating under normal condition by the sound of a mixer. Some companies use a stethoscope to make sure nothing unusual happens inside the mixer.

(2) Coating process

The coating machine consists of coating head, dry zone and roll driving part. Among various coating methods, comma roll and slit die coating methods are most commonly used in the battery manufacturing. A comma roll coating method can control the thickness of the electrode accurately whereas the slit die coating method is faster than the comma roll coating with a little sacrifice of the accuracy. LG Electronics, a sister company of LG Chemical, had produced video tape using the comma roll coating method, so LG was familiar with comma roll coating. In contrast, Samsung didn't participate in the video tape business, so they didn't have any hands-on experience in the coating technology. LG Chemical chose the comma roll coater of Hirano of Japan for the purpose of utilizing the accumulated experience associated with comma roll coating, and

Samsung SDI chose the slit die coater of Toray of Japan for the high speed.

As regards the drying method, most battery makers adopted an overhead nozzle type drying method, in which dry air is sprayed on the coating film. With regard to the dry zone, each zone has its own role. The first and second dry zones determine the position of the binder. The third zone fixes the binder position and the fourth dry zone stabilizes the electrodes. The fundamental purpose of the coating process is to distribute the binder uniformly in the electrode. The dry zones were designed to serve this purpose.

When we use the water based binder, temperature of the first dry zone is important in the electrode quality. It is advisable that we maintain the temperature below 60°C in the first dry zone. Otherwise, CMC, a surface active agent, would float to the surface of the film, which results in a serious binding problem of the electrode. Intermittent coating is a pattern coating in which the coated and uncoated areas are alternated. The quality of the pattern coating is determined by the sharpness of the boundary between coated and uncoated area. The electrode thickness after coating is representative of the loading level, which is the design parameter of the cell. The maximum loading level depends upon the characteristics of the active materials. Since lithium ion batteries do not have a lithium source, the ratio of the anode to cathode capacity becomes an important design parameter, which is called the n/p ratio. Battery makers stipulate the minimum n/p ratio in their design standard to prevent the haphazard cell design, which more often than not culminates in the field incidents. Many battery makers appoint 1.12 as a reasonable minimum n/p ratio in their cell design rule. A low n/p ratio increases the energy capacity at the expense of the safety. On the contrary, a high n/p ratio increases cycle life with the decrease of energy capacity. In case the n/p ratio is below 1.06, cells are vulnerable to the outbreak of fire and explosion unless the electrode thickness is uniform, as lithium metal is precipitated on the anode surface because of the local n/p ratio reversal. In actuality, the batteries that caused the field incidents exhibited very low n/p ratio, mostly below 1.06.

Aluminum and copper foil is the current collector that collects electrons from the electrode and transfer electrons along the electron path. Metal foil also acts as the substrate for coating process. Both sides of the foil are coated with electrode slurry. One side of the foil is coated with electrode slurry and then the coated roll is returned to the coating head for the coating of the opposite side. As the foil is moved horizontally in the drying zone, we cannot coat the slurry on both sides of the foil simultaneously. Thickness of the electrode is measured just before winder, and this information is fed back to the coating head for the closed loop feedback control purpose. Any slight vibration of the coating web can have influence on the coating condition, so it is very important that we maintain the calm environment in coating.

(3) Roll pressing and slitting

Rolls of a rolling machine consist of working rolls and back-up rolls. Working rolls actually press the electrodes down while the back-up rolls support the load applied to the working rolls. Number of back-up rolls determines the power and accuracy of thickness. The more the back-up rolls, the more expensive the rolling machine is. In a lithium ion battery, 2-high mill is generally used, which is composed of one pair of working rolls and back-up rolls respectively. Elevated temperature roll pressing is possible if roll heating system is installed, but in most cases cold rolling is preferred for the accurate control of thickness.

Roll pressing had been introduced to enhance the cell performance further. First, electrode density is increased by reducing the electrode thickness. Second, binding force is improved between the cathode and aluminum foil. Aluminum exhibits notoriously poor binding force on account of the aluminum oxide film on the surface of aluminum. Rolling force breaks the aluminum oxide film, which induces mechanical adhesion by mechanical interlocking of two surfaces. Third, crystallographic texture introduced by rolling improves the power capability of the cell. When rolling pressure is applied to the surface of the electrodes, active materials align themselves in a given direction, the phenomenon of which is called texturing. Textured electrode facilitates the movement of lithium ions in the electrode.

Electrode density, which is an important design parameter, is determined after roll pressing. In small lithium ion batteries used in mobile phones and laptop computers, competition had been centered on increasing the running time. The design concept for the implementation of the increase in running time is simple and clear. Increase in the amount of active materials with a concomitant decrease in the amount of inert materials like a separator and the current collector without any influence on the safety will increase the energy capacity of the battery. However, it should be noted that high energy density has a harmful effect on the safety. If the electrode density is too high, batteries will spew out the jelly roll and electrolytes in case of CID opening, which makes the batteries susceptible to serious safety incidents.

In an automotive application, batteries should last for over ten years. In order to maintain the cycle life and storage life to this level, it is essential not to increase the electrode density over a certain level. Excessive electrode density causes wrinkles on uncoated foil in roll pressing. Low electrode density is also beneficial to the cycle life as it responds to the stress properly and holds more electrolytes.

After roll pressing, we cut the rolled electrode into many rolls with small width corresponding to the size of the jelly-roll, which is called a slitting process. The slitter consists of blades and a winding unit. The winding unit has a big unwinder and several small winders. We refer to rolled electrodes made by a slitting process as the jumbo roll.

3) Battery assembly process

A vacuum drying process removes water included unintentionally during electrode making process. Jumbo roll is dried at temperatures above water vaporization temperature for over 12 hours. In a slitting process, winding tension of the jumbo roll determines the subsequent behavior of the jumbo roll since this tension is embedded in the jumbo roll in the form of residual stress. During a vacuum drying process, a jumbo roll may deteriorate if the residual stress in the jumbo roll is not appropriate. In designing manufacturing methodology, it is advisable to consider the slitting, vacuum drying and winding process as a single unit because each process more or less affects the subsequent process.

In prismatic batteries, winding process is classified into a flat winding and a diamond shape winding depending on the mandrel design. In using a diamond shape mandrel, a jelly roll is pressed after winding to make a flat jelly roll. Tension applied to the electrode in winding has influence on the cycle life since residual stress embedded by applied tension determines deformation pattern in charging and discharging. In cylindrical batteries, the anode tab is welded to the bottom of the metal can and cathode tab is welded to the top cap. Thus, the top cap becomes the positive terminal and the metal can becomes the negative terminal. On the contrary, in prismatic batteries, the cathode tab is welded to the side wall of the inside of the aluminum can and the anode tab is welded to the top cap. Thus, the metal can becomes the positive terminal and the top cap becomes the negative terminal.

The metal cans are supplied from can makers, but responsibility for the development of a can is primarily on battery companies. Metal cans are made by a deep drawing process of the metal sheet. Deep drawing is the metal forming process in which a metal sheet is pressed on the forming die to make a cup shaped can. The maximum height of the cup formed by a deep drawing process is same as the diameter. Therefore, to make a cylindrical can used in the battery, it is required to go through a series of deep drawing, which is called a progressive stamping. In prismatic batteries, the cross section of the cup changes from circle to oval type to finally rectangular shape in the progressive stamping process.

Electrodes and the separator of nickel metal hydride batteries are thick and have many pores. Therefore, centrifugal force is applied to distribute the electrolyte uniformly after an electrolyte injection. On the other hand, the electrodes of lithium ion batteries are thin and dense, so centrifugal force is not strong enough to distribute the electrolyte inside the pores of electrodes and a separator. In lithium ion batteries, pressure should be applied to the injected electrolyte for a certain period of time to push the electrolyte into the dense electrodes by force. First, remove the air inside the metal can by vacuuming out a gas and then inject the electrolyte into the can. After the electrolyte injection, apply the pressure to impregnate the electrolyte in the pores of the electrodes and a separator by force. In all the sealed type batteries, the amount of electrolytes plays an important role in determining the state of health (SOH) of the battery regardless of the battery system. The HIBAR pump is used to inject an exact amount of electrolytes in lithium ion batteries. The leading battery firms which are keen to produce uniform cells manage a HIBAR pump throughly to make sure that they control the amount of electrolytes injected into the cells accurately.

One of the core technologies in the battery manufacturing process is the welding technology. Various kinds of welding techniques had been employed in a battery manufacturing process such as laser welding, ultrasonic welding and resistance welding. Some battery companies consider adopting more advanced welding technology like an electron beam welding in the manufacturing process. As in coating process, vibration can affect the quality of welding, so the operating condition should be closely controlled not to be affected by the vibration.

There are two impedance measurements adopted in the battery manufacturing process: DC-IR and AC-IR. The purpose of the DC-IR measurement is to evaluate the soundness of the electron path while the AC-IR assesses the status of the electrolyte channels. Combination of both measurements gives us a clear picture on the overall soundness of the battery.

4) Formation and aging process

The design philosophy of lithium ion battery manufacturing is to enable us to track down each and every cell which has been produced and shipped out. Formation process acts as the control center of the battery production for the pursuit of back-tracking of cells among other things. In effect, it is the data management system of the formation process that determines whether or not the battery maker is ready to enter the major battery market. Hence, their data management capability including the back-tracking of cells is evaluated in the auditing of the battery makers.

An accurate amount of electricity should be charged to each cell in the formation process. Inaccuracy in the amount of electricity charged to the cells changes the thickness of the solid electrolyte inter-phase (SEI) film on the anode, which results in the deviation of the energy capacity of the cells in the same batch. Therefore, the quality of the formation process is determined by how accurately the electricity is infused into the cell. To control the accuracy of the formation process, terminals of cells and charging equipment should be free of contaminants.

Another factor that has influence on the distribution of energy capacity of each cell is temperature. Temperature is the most dominant factor in the formation process and the cell performance. Unless the temperature distribution of the pallet is uniform, energy capacity of cells in the pallet will be different from each other even though we control the amount of charged electricity of each cell accurately. When the cells are moved from the production to the formation building by an automated guided vehicle (AGV), pallets with cells inserted in sequence are moved to the formation room and the empty pallets are returned to the production building. During this transport process, pallets can be slightly distorted by mishandling. This minor dimensional change of the pallets can have an effect on the quality of cells by inducing non-uniform distribution of cell temperatures across the pallet.

Aging period is called the qualification period in that it eliminates not only defective cells but any suspicious cells that may cause problems in the field. First, hard short cells are sorted out. Leave activated cells at a certain temperature and humidity for 48 hours, and measure the voltage and impedance to screen out the defective cells. After hard short screening, soft short cells are sorted out. Leave cells at a certain temperature and humidity for 16 days, and measure the voltage and impedance. Based on the voltage and impedance, we sort out the defective cells.

Both hard and soft shorts result from the accidental inclusion of metallic impurities during manufacturing process. It should be noted that the design of the formation system is based on the concept that the only defects that can appear in lithium ion batteries are hard and soft shorts. In case of the appearance of other defects than hard and soft shorts, the formation system designed to sort out defective cells cannot be applied. The primary criterion for the eligibility for mass production of lithium ion batteries is whether or not the defects of cells in the formation process are limited to hard and soft shorts caused by the accidental inclusion of copper, nickel and iron. If other defects than hard and soft shorts appear frequently, it is strongly recommended that you go back to the R&D stage. However, more often than not, this useful criterion in the formation process as to the readiness of mass production is completely ignored.

The water-based binder used in natural graphite generates a large amount of gas in the formation process, so precharging process is adopted to eliminate the gas. The electrolyte is injected by two stages to prevent the overflow. Precharging is conducted before sealing the cell, so it is considered as a part of the assembly process and is carried out in the dry room of the battery assembly building. First, 10% of the electrolyte is injected to the battery, and charge the battery for six minutes at 1C rate for the achievement of 10% state of charge (SOC), followed by 20% electrolyte injection and charging for twelve minutes at 1C rate. SOC is increased to 30% after the precharging process. Subsequent to the injection of the rest 70% electrolyte, the battery is sealed and moved to the formation room. After formation, leave the battery at 65°C, which is called high temperature aging. High temperature aging stabilizes a SEI film and enhances the uniformity with great efficiency.

Cells can be burnt or exploded in the aging process. We call this incident an event in the lithium ion battery business. According to the rule book of lithium ion battery operation, we are to discard all the cells corresponding to the batch of cells that caused the event, and to stop the production line until the cause of the event is identified. Therefore, once the event takes place in the battery plant, the damage is enormous in terms of the money and time. That's why it is called a poison to the lithium ion battery business. Nevertheless, it is the rule which all the lithium ion battery makers should follow to insure the safety of cells because investigation shows that most of the cells that caused the field incidents like fire or explosion of the battery belong to the batch in which some of cells had burnt or exploded in the aging process.

07
:
Management of batteries

1) Degradation mechanism

Comprehensive understanding of how people are getting old would enable us to find a way to stay healthy and live a long life. Similarly, knowledge on how the battery gets degraded will give us the information on how to develop a battery with a long cycle life. The battery gets degraded in three steps: a jelly roll deformation, destruction of the electrolyte channels and the electrolyte dry-out.

The battery changes its volume in charging and discharging, which can be compared to breathing. Volume change is a natural process in the battery, but the way the volume changes determines whether or not it would lead to permanent plastic deformation. Deformation mode of the jell roll is affected by the residual stress embedded in the winding process and the uniformity of the electrode thickness. Accurately controlled winding process along with uniform thickness of electrodes ensures that the battery maintains the energy capacity with cycling by inducing reversible volume change. Therefore, the most convincing way to assess battery company's capability is to compare the configuration of the jelly roll cycled to 50, 100 and 200 cycles with that of fresh cells in terms of the robustness and configuration change of the jelly roll. The jelly roll made by the leading battery companies with a great deal of experience maintains the soundness even after over 200 cycles. In contrast, the jelly roll made by inexperienced companies exhibits the sign of degradation in the first 10 cycles. In this respect, we can say that experience and knowledge of battery makers are reflected in the rigidity of the jelly roll.

The onset of the abrupt decline of the cycle life curve coincides with the inception of the second stage degradation, which is the destruction of the electrolyte channels. The solid electrolyte inter-phase (SEI) film is formed by the chemical reaction between graphite and the electrolyte, and a small amount of gas is generated during formation process. Since the empty space in the can accommodates these gases, the accurate volume of the free space inside the can should be specified in the design of the cell. As the cycle progresses, the SEI film is damaged and destroyed locally. Local destruction of the SEI film is due to the non-uniform flow of lithium ions. If lithium ion flow is concentrated on any locality, which happens under the condition of high current, the SEI film is broken locally and naked graphite is exposed to the electrolyte. Exposure of the naked graphite to the electrolyte brings about reaction between graphite and the electrolyte, and forms a fresh SEI film locally at the expense of the electrolyte, during which gas is also generated. As this process continues, the electrolyte channel is disrupted, which isolates some of active materials. Isolated active materials no longer contribute to generating the electricity. Isolated active materials in combination with the reduced electrolyte channel by the loss of the electrolyte for the formation of a fresh SEI film reduce the energy capacity with cycles, which is the mechanism of the second stage degradation. Isolation of active materials also occurs in the cathode.

Transition from the first to the second stage of degradation is related to the fatigue of copper and aluminum foil. Metal is subject to the fatigue failure. Fatigue is a form of metal fracture in which repetitive application of the stress lower than the yield strength leads to the breakage due to the accumulation of the localized plastic deformation. At the end of the first stage, metal foil undergoes permanent plastic deformation caused by fatigue, which induces non-uniform lithium ion flows. Fatigue is unavoidable as long as we use the metal foil as the current collector.

Battery is dead when the battery is running out of the electrolyte, which is called a dry-out. Secondary batteries are not usually used until the electrolyte dries out. The battery's life is defined as the time span from the beginning of life (BOL) to the end of life (EOL). In mobile phones and laptop computers, the EOL is defined as the number of cycles at which 80% of the energy capacity is retained. Even when the EOL is reached in the mobile phones and laptop computers, they can be operated without any serious problems, but users are beginning to feel that running time got diminished. In the age of nickel cadmium batteries and in the early period of advanced batteries like nickel metal hydride and lithium ion batteries, the EOL was defined as the 60% charge retention. However, as the energy density of advanced batteries increases, portable electronics companies increased the level of cycle life by increasing the EOL from 60% to 80% charge retention.

In the hybrid electric vehicles, power is the main design parameter rather than the energy capacity, and the duty cycle, which reflects the driving pattern of the vehicle, has influence on the durability of the battery. In general, 60% power retention after 300,000 cycles of the duty cycle of hybrid electric vehicles is defined as the EOL of the battery.

2) Field incidents

Prior to the introduction of lithium ion batteries in the portable electronics market, lithium secondary batteries adopting the lithium metal anode entered the portable electronics market. Battery scientists and engineers with good battery technology background gathered together and formed a battery company to commercialize lithium secondary batteries with high voltage, high energy capacity and a long cycle life. This lithium battery was considered to be unbeatable at the time when a conventional nickel cadmium battery was a major battery in the portable electronics market. It was Moli Energy of Canada.

Moli Energy's lithium secondary batteries were installed in NEC's mobile phones in Japan and were in good use for some time with mobile phone users' full satisfaction with the extended running time. One day, however, one of Moli Energy's lithium cells was reported to have created field incident during usage at the NEC's mobile phone. What had actually happened was that the vent was opened with a flame, but this event gave rise to malicious rumors including the rumor about the serious injury of a mobile phone user. These rumors, despite Moli Energy's denial, persisted, which led to the demise of Moli Energy, culminating in NEC's acquisition of Moli Energy.

Regardless of the investigation results on what had caused the field incident of Moli Energy's cells and whether or not this phenomenon would be rectifiable, the lithium secondary battery, in spite of its obvious advantage of high energy capacity, was forced to leave the market, and Sony's lithium ion battery filled in the position which Moli Energy's lithium secondary battery had held. Sony was keen to stress that lithium ion batteries don't catch on fire or explode since lithium exists not in the form of atoms but in a form of ions. Sony's focus on safety not to repeat the same mistake of Moli Energy reinforced its capability in the markets it had served.

Sony's belief that the lithium ion battery does not explode under any circumstances lasted for ten years without any noticeable incidents.

However, Sony's belief crumbled in 2002-2006 with the outbreak of field incidents under the normal operating condition all over the world. The question naturally arises as to what made lithium ion batteries catch fire or explode and, more importantly, why the field incidents took place after the elapse of ten years. These questions should be answered to prevent the recurrence of field incidents we had experienced in 2002-2006.

One of the findings in relation to the field incidents was that it was beginning to take place when the energy capacity of 18650 cylindrical batteries reached 2Ah, the acceptance of which implies that the energy density might be a decisive factor in the field incident. Increase in the energy density makes it difficult to design the battery without any weak areas through which energy leaks out uncontrollably. In the lithium ion battery which uses graphite and LCO as the anode and cathode materials respectively, the maximum energy capacity of 18650 cylindrical batteries that can be made without any risk of safety issues is below 2Ah. When the energy capacity exceeds this maximum limit, no matter how carefully the cell is designed, chances are there will be several weak areas from which energy leaks out. By and large, that's the way field incidents happened and, more importantly, that's the reason why the field incidents broke out after the elapse of ten years.

Battery makers have to know how to cope with field incidents. Once the field incident is reported to the mobile phones or laptop computer makers, the burnt power pack is collected from the site where field incident had happened by portable electronics maker for the investigation of what had caused the incident. If the investigation results indicate that the field incident may have been caused by the change of cell design or materials, all the cells and portable electronics related to the change of design should be recalled on account of the strong likelihood of the continuous occurrence of field incidents at haphazard. This would damage the battery maker as well as portable electronics maker seriously, the extreme case of which is found in the Moli Energy's lithium secondary batteries. On the other hand, in case the field incident is caused by the operator's mistake in manufacturing such as the inclusion of metallic impurities by accident, the battery maker will be able to minimize the damage by shipping back cells relevant to the specific batch. Nonetheless, a scar left by the field incident is so deep that it would take at least a year for the battery maker to get over the impact and the aftermath caused by the management shakeup accruing from the field incident.

For the determination of the cause of fire, the first thing that should be done is to find the origin of fire. The first step is to spread out the electrodes of the burnt cell on the table and look for the origin of ignition. In most cases, the ignition site is located at the edge of electrodes where the coated materials were torn off or cracked. Defective edges of the electrodes make the separator practically sandwiched between copper and aluminum foil without any presence of electrode materials, which enables the cell to be very susceptible to hard short, and is apt to increase the temperature high enough to ignite the electrodes.

However, if the ignition site is located in other areas than the edges of the electrodes, the damaged separator caused by the forced insertion of the jelly-roll in prismatic batteries or the forced assembly of pouch batteries may be the cause of fire or explosion. Forced insertion of the jelly-roll inside the can or the forced assembly of the pouch case results from the increased volume of the jelly-roll and the concomitant reduction of the free space inside the can or the pouch case. In summary, the presence of the ignition site in the location other than the edges of the electrode strongly indicates that the fire or explosion of the battery broke out because energy density of the battery is higher than the maximum upper limit of energy density. The complicated problem associated with the maximum energy density of the battery is that we cannot estimate or predict the maximum energy density in advance. Only when the field incident breaks out, we will be able to gain an understanding of the maximum energy density. Therefore, more often than not, the leading maker is prone to become the first victim. The leading maker is the one that puts the battery of the highest energy density on the market. In laptop computers of 2002-2006, Sanyo and Dell, the leading battery and laptop computer maker, became the first victim of the filed incident in 2002.

Whether or not, and how, the fire spreads out depend upon how robustly the cell has been designed and made. For example, the tightly wound jelly roll made with high density electrodes propagates the flame so expeditiously that it can easily lead to fire of cells or explosion. Active materials with high safety features such as spinel manganese oxide (LMO) or lithium iron phosphate (LFP) are likely to alleviate the spreading power of flame, if not completely blocking the fire spread. The amount of free electrolytes has influence on this process in that free electrolytes facilitate or accelerate the propagation of the flame.

The pack design also determines how severely the field event can be developed after initial flame propagation stage. Configuration of cells and PCM should be designed not to disrupt the natural flow of a gas and the electrolyte. If the electrolyte which is flammable is sprayed on the hot circuit board, PCM is in danger of catching fire. When the CID is opened under overcharging condition, only the gas, mostly carbon dioxide, is supposed to be released. But if electrode density is too high or free electrolyte is too abundant, the cell is likely to spew out the hot electrolyte and a jelly roll along with a gas. If spewed-out electrolyte spreads on the hot circuit board, it will increase the likelihood of developing into fire.

When we back-track the cells that caused field incidents, in most cases the batch of the cells coincides with the batch in which some cells have exhibited a warning signal like fire or explosion in formation and aging process. In this regard, we can cautiously conclude that the field incident can be prevented by improving the battery production management. Sometimes the event of fire or explosion takes place during formation and aging process in the battery plant. In case that happens, we have to take an immediate measure of stopping battery production and discarding all the cells of the batch burnt or exploded cells belong to, and then investigate what had caused the incident. Only after the problem is rectified, we are allowed to restart the equipment. This screening system was originally designed by Sony not to ship out any suspicious cells that may cause field incidents during usage. However, it takes so long to investigate the cause of the incidents that some battery companies prefer to opt for less expensive, but highly risky alternatives. Instead of stopping the production line, they tend to have cells of that batch undergo several charging and discharging cycles to simulate the situation of battery usage. In case no event of fire or explosion takes place during cycling, they ship out the cells of that batch. Cells shipped out through this unorthodox screening process often result in field incidents.

In the battery business, it is of paramount importance for managers to have control over any issues that might have association with the safety, however minute. The following example gives us a case in point. Electron conducting materials like carbon black that is added in the cathode to improve the conductivity of electrodes is classified as the minor material because it is used in small quantity and is believed to have no influence on the safety. With this in mind, a manager of the battery plant changed the vendor of carbon black without going through the process related to the adoption of a new supplier. The cells made with new carbon black broke out a series of field incidents. It took them several months to figure out what had caused field incidents. New carbon black introduced by a manager with the intention of cutting some corners nearly demolished the battery business. In sum, strange as it may sound, management is a key area where we'll have to make improvement to prevent field incidents.

Field incidents in 2002-2006 changed the competition landscape dramatically. Dell recalled lap top computers powered by batteries made by Sony and Sanyo and Apple also recalled lap top computers using LG Chemical's batteries. As a result of recall, Sanyo was merged to Panasonic in December, 2009. Sony seriously considered giving up on the battery business. Sony had suffered from the aftermath of field incidents, and finally sold the battery business to Murata in 2016. LG Chemical struggled heavily after field incidents and the recall, but they managed to go through with a difficult time and recovered from the aftermath of field incidents by signing the development contract with GM in 2007. As regards Samsung SDI, HP decided not to recall lap top computers in spite of the fact that field incidents continued to take place for six months in 2003. Thanks to HP's strategic decision not to recall lap top computers, Samsung SDI became a lone survivor. Samsung SDI reached the top position in 2011 and had maintained the top position until 2014.

3) Field incidents of Galaxy Note 7

Explosion of the smart phones that took place in August, 2016 flabbergasted the battery industry because no conspicuous field incident took place for the period of ten years. Galaxy Note 7 of Samsung Electronics powered by Samsung SDI's lithium ion batteries with the energy capacity of 3,500mAh started to explode in 5 days after launching of the product. When the seventh explosion of Galaxy Note 7 took place in the field, Samsung Electronics decided to prohibit selling its products and started to recall suspicious smart phones. In retrospect, Samsung Electronics irrationally rushed into launching Galaxy Note 7 in order to introduce Galaxy Note 7 ahead of Apple's iPhone 7 and iPhone 7 Plus which were scheduled to be put in the market on September 7, 2016.

After the field incidents in 2002-2006, the battery industry changed the cathode material from LCO to NCA and NCM which have higher specific capacity than LCO. Safety of the battery had been maintained for ten years because of the use of nickel base cathode materials. Samsung SDI increased the energy capacity of the battery from 3,000mAh to 3,500mAh to extend the running time without use of NCA or NCM. Interestingly enough, Samsung SDI used LCO in the cathode instead of NCA or NCM. LG V20 of LG Electronics, which is same as Galaxy Note 7 in size and has a more powerful audio function, uses the battery whose energy capacity is 3,200mAh. The battery capacity of Apple's iPhone 7 Plus is only 2,900mAh. Considering the excessive increase of the energy capacity of the battery, the high energy density may be the root cause of the explosion. Lithium ion batteries explode not only in the abused condition but also in the normal operating condition if the energy density is higher than the maximum energy density.

The pouch case of the battery is fabricated by pressing the aluminum sealing paper on the die. The maximum thickness of the pouch case we can make by a pressing process is 8mm, but the practical maximum thickness is 6mm, above which the jelly-roll of the battery which consists of electrodes and a separator can be damaged in the assembly process. The thickness of the embedded pouch battery of Galaxy Note 7 is 7.9mm.

In order to make a battery with the energy capacity of 3,500mAh, we should increase the volume of the jelly-roll with the concomitant reduction of the free space inside the case. Needless to say, the increased volume of the jelly-roll makes it difficult to assemble a jelly-roll inside the pouch case, which might have made the jelly-roll susceptible to damage in the assembly process. Some batteries may have had the puncture in the separator due to the forced assembly of the cell. Puncture of the separator results in the short by direct contact of the anode and cathode. The short increases the temperature locally, and the heat spreads to the electrodes fast. When the temperature increases to the ignition temperature of the electrolyte, the electrolyte catches on fire. Once the electrolyte catches fire, the battery spew out the jelly-roll and the electrolyte at high speed. The hot electrolyte is sprayed on the hot PCM with a flame and fire spreads to the pack and the main body of the smart phone.

All the field incidents start with the internal short, but whether or not the short leads to fire or explosion depends on the design and the energy density of the battery. In Galaxy Note 7, explosion was powerful enough to burn the display of the smart phone. Samsung SDI reduced the volume of the inert materials and increased the amount of active materials to meet the requirement of the high energy density. The thickness of the separator was reduced drastically, so any defects of the separator or the slight mishandling of the separator in the battery manufacturing process could lead to the internal short. Samsung SDI also increased the electrode density excessively, so the electrodes are filled with concentrated energy.

When the cell is designed according to the design rule, the accidental short by the puncture of a separator does not lead to fire or explosion. Instead, the battery is dead by the operation of the vent. The accidental short increases the temperature and simultaneously the internal pressure of the cell is increased by the formation of gas. The gas is released through the vent, as a result of which the battery is dead. Sometimes smoke can be formed when the battery is dead. Development of explosion from the short is a rare occurrence in the battery. The basic principle of the battery design is to prevent the short from being developed to explosion. However, the accidental short of the battery was developed into explosion due to the concentrated energy in the electrodes in Galaxy Note 7. Samsung SDI may have ignored the critical safety factors such as n/p ratio, the free space inside the case, the maximum electrode density, the amount of free electrolyte, the minimum separator thickness and the maximum thickness of the pouch battery to reach the energy capacity target of 3,500mAh.

As regarding Samsung SDI, the field incident took place thirteen years after the explosion of HP's laptop computers in 2003. Ironically, the design methodology used in increasing the energy density of laptop computers and smart phones is identical. One of the key technologies associated with the increase of the energy density of Samsung SDI is the concentration of the energy in the electrodes, and this technology seems to have been maintained since 2000. Because of the concentration of the energy in the electrodes, Samsung SDI's batteries tend to dissipate high explosive power as compared with batteries of other battery companies in the field incidents.

The energy capacity of the battery used in Galaxy Note 7 is higher than that of batteries used in LG V20 and Apple's iPhone 7 Plus by 300-600mAh. The basic concept of the competition in the energy capacity for the extended running time is to take one step ahead of others in the energy capacity, so the difference of 100-200mAh is the normal spectrum of the energy capacity in the battery used in the same market. Development of the battery whose energy capacity is higher than that of the competitors' batteries by 300-600mAh without introduction of new materials implies that something unreasonable must have been implemented in the design of batteries.

Samsung SDI used a separator whose thickness is 8 micrometers. The thickness of the separator that caused the filed incident in HP's laptop computers in 2003 was 18 micrometers, which means that an extremely thin separator was used in the pouch battery of Galaxy Note 7. When the extremely thin separator is used, the separator fails to separate the anode and cathode electrically. Microcurent flows between the anode and cathode, the phenomenon of which is called the soft short. A charger recognizes the charging termination voltage based on the current after reaching the charging voltage, which is 4.2V in the LCO battery. Microcurrent due to the soft short deteriorates the charging algorithm, and the charger keeps charging the battery after reaching the full capacity. That's why ill-designed batteries are more vulnerable to fire or explosion in charging than in discharging. In the batteries of Galaxy Note 7, charging voltage increased beyond the charging termination voltage due to the soft short. Overcharging must have taken place in the battery frequently because of the difficulty to control the charging voltage. Therefore, the soft short in addition to the hard short may have contributed to the initiation of the field incident.

Nickel base cathode materials, NCA and NCM, have the secondary structure. Small particles agglomerate to form a main particle in the secondary structure. Because of the powder morphology of NCA and NCM, it's hard to increase the electrode density above 3.2g/cc. In contrast to NCA and NCM, LCO has the primary structure, so it is amenable to being densified by roll pressing. One of the core technologies of Samsung SDI is to increase the density of the electrode. In order to take advantage of its electrodes densification technology, Samsung SDI used LCO instead of NCA and NCM in the cathode and increased the electrode density to 4.2-4.4g/cc. When Samsung SDI was the victim of field incidents in HP's laptop computers in 2003, the electrode density of LCO cathode was 3.2g/cc. Samsung SDI must have increased the electrode density dangerously high to meet the requirement of energy capacity of 3,500mAh. Concentrated energy in the electrodes must have dissipated the explosive power in Galaxy Note 7.

Considering that the density of LCO is 5.1g/cc, the fact that the electrode density is 4.4g/cc implies that the space for the electrolyte channel in the electrode is limited. In addition, the high electrode density makes it extremely difficult for the electrolyte to penetrate into the pores of the electrode. Therefore, as time goes by, the Samsung SDI's pouch batteries for Galaxy Note 7 may be in danger of the electrolyte separation due to the incomplete formation of the electrolyte channel. The electrolyte inside the electrode seeps out to form free electrolytes, which is called the electrolyte separation. The electrolyte separation is also the cause of the limited cycle life and swelling.

Samsung Electronics decided to terminate the production of Galaxy Note 7 in October, 2016, 52 days after the launching of the new product. Repetitive explosion of the replaced phones forced Samsung Electronics to discontinue the production of Galaxy Note 7. Replaced phones were susceptible to explosion by slight mechanical impact, which is a typical phenomenon of electrolyte separation due to the unstable electrolyte channel.

08
:
Lithium ion battery makers

1) Japanese companies

Up until the 1980s, the world battery industry was dominated by western nations. Dry cells and lead acid batteries, the first primary and secondary batteries, were invented by French scientists in mid 1800s. The nickel cadmium battery, the first advanced secondary battery, was invented in Sweden and was commercialized by Varta of Germany and SAFT of France. As regarding the primary battery, Eveready and Duracell of the USA have been leading the world alkaline manganese battery market for a long time.

As compared with other industries of Japan, Japanese battery industry was rather weak before World War II. German U-boat anchored along a pier of Osaka harbor was deserted by German crew because of the termination of World War II in 1945. Japanese engineers from Panasonic had found nickel cadmium batteries inside the U-boat, and they investigated the nickel cadmium battery by reverse engineering. Based on the information collected from the analysis of the German nickel cadmium battery, Panasonic managed to figure out how to make nickel cadmium battery, which was practically the starting point of the secondary battery business of Japan.[1]

The Tokyo Olympic Games held in 1964 were the important mile-stone for the road to the heyday of Japan. In the 1960s electronic industry of Japan regained its strength to compete with western countries. In response to the demand of electronic industry, Sanyo developed small nickel cadmium batteries. It developed the sealed type nickel cadmium batteries by taking advantage of the oxygen recombination mechanism that prevents the evolution of the gas in overcharging. Sanyo's development of small nickel cadmium batteries expanded the area of the secondary batteries from automotive and industrial applications to portable electronics. In the 1970s, Sanyo and Panasonic developed lithium primary batteries, which doubled the operating voltage from 1.5V to 3V. Lithium primary batteries were used in the automatic camera all over the world.

Tight competition between Japan and western countries started in the early 1980s in the area of automobiles and electronic appliances. Japanese influence on the US automotive industry was so strong that the US automotive designers had to accommodate their design concept to the changing environment. As a result, automobiles made in the USA had the tendency to be smaller than they used to be.

(1) Battery development competition

World economy of the 1980s showed a clear sign of recovery after the trauma caused by the oil shock of 1973 and 1979. One of the emerging markets of the 1980s was portable electronics. Supply of integrated circuit (IC) chips with reasonable price by semiconductor industry facilitated the development of portable electronics. Since the power source of portable electronics is the secondary batteries, the necessity of high performance advanced secondary batteries naturally arose in the 1980s. The only small

1 New Battery Technologies, Matsushita Battery Inc., 1994.

secondary battery from which electronic firms could choose at that time was the nickel cadmium battery developed by Sanyo of Japan in 1960. Needless to say, lack of choice in the secondary battery motivated the development of the advanced secondary batteries. There was a competition for the development of advanced secondary batteries between Europe and North America and Japan in the 1980s, so the 1980s was recorded as the decade of the competition of the advanced battery development. Japanese battery makers, the winner of the battery development competition, have been leading the world battery industry.

The insight which European companies had provided relative to their target battery certainly defined a course which would be the most challenging. Their target product was a LPB, in which a polymer electrolyte acts as both a separator and the electrolyte simultaneously. The Anglo-Danish and French-Canadian project funded by European Community had led the development of LPB. The Anglo-Danish project had taken great strides in its progress. In effect, development results of an Anglo-Danish project became the foundation on the commercialization activities of Valence Technology of the USA.

North America took a more practical step than Europe. Moli Energy of Canada developed lithium secondary batteries incorporated with liquid electrolytes and entered the portable electronics market in 1988. This battery outperformed the nickel cadmium battery in all the performance categories including cell voltage, energy density and cycle life with the exception of power capability. It is dominance in the portable electronics market looked so imminent and obvious.

A remarkable increase in the interest of lithium ion batteries occurred in Japan probably inspired by a series of success in the intercalation of lithium atoms to carbon in the 1980s. Another concern of Japan was to find out the way to eliminate cadmium metal in the nickel cadmium battery to solve the problems associated with cadmium: itai itai disease and the memory effect. Hence, Japanese battery makers focused on more realistic targets: nickel metal hydride and lithium ion batteries.

The news of the field incidents of Moli Energy's lithium secondary batteries flabbergasted many people in the battery industry. At that time, we never thought that the batteries could actually catch fire or explode in a real environment. NEC decided to recall the mobile phones because of the fire risk in 1989. Recall of NEC's mobile phones powered by lithium secondary batteries made by Moli Energy of Canada was a sign of the demise of a lithium secondary battery incorporated with lithium metal anode.

Valence Technology which had commercialized a LPB based on the result of Anglo-Danish project failed to meet the requirements of Motorola's mobile phones in the early 1990s. The common denominator of a LPB of Valence Technology and a lithium secondary battery of Moli Energy was that they all adopted the lithium metal anode. Therefore, their setback made people have a second thought on the use of lithium metal anode in the secondary battery. Lithium metal anode is still not used in the secondary batteries despite the long development history.

Among four battery systems developed in the 1980s, only nickel metal hydride and lithium ion batteries were successful in commercialization and have been maintaining the viability in the battery industry. With the success of new advanced secondary batteries developed in Japan, Japanese battery companies emerged as the leading battery makers in the world battery market.

In retrospect, the 1980s is considered not only as the era of new battery development but also as the era when Japanese dominance started to manifest itself.

(2) Major battery makers

Up until Korean companies, LG Chemical and Samsung SDI, entered the lithium ion battery market in the late 1990s, lithium ion batteries were produced solely by Japanese makers. In this respect, the lithium ion battery had been naturally called the Japanese battery for quite a long time. Sony, Sanyo and Panasonic formed the leading group in lithium ion battery industry, followed by ATB (Asahi Toshiba Battery), GS-Melcotec, NEC-Moli and Hitachi-Maxell. These seven Japanese battery makers formed a formidable force and dominated the world battery market throughout the 1990s.

Sony successfully commercialized lithium ion batteries in 1991. Sony became the first company to open the door of lithium ion battery market. The next battery makers that joined in the lithium ion battery market were Sanyo and Panasonic. Sanyo was strong in the core technology like battery materials and cell design. Panasonic had the advantages over its competitors in the area of battery manufacturing technology. Sanyo and Panasonic along with Sony had enjoyed the position of major battery makers for a long time before Korean battery makers disrupted competitive composition in 2000, almost ten years after the inception of the production of lithium ion batteries.

The battery industry owes the current status of secondary batteries to Sony. Sony designed the manufacturing technology of lithium ion batteries. In addition, it developed the mass production technology of LCO cathode materials and designed the PCM and the CID among other things. Laptop computer makers of the 1990s wanted to change the batteries from nickel metal hydride to lithium ion batteries because the nickel metal hydride battery had poor high temperature performance. However, the concern on the safety of lithium ion batteries in the multi-cell application made the laptop computer makers hesitate to adopt lithium ion batteries. Battery makers alleviated the safety concern of laptop computer makers by the combined adoption of the CID and fuel gauge. Fuel gauge is the energy capacity measuring device. Thanks to these safety devices, the lithium ion battery drove away the nickel metal hydride battery from the laptop computer market.

Sony's another important accomplishment is the development of the pouch battery which uses a thin separator instead of a polymer electrolyte to have the competitive edge over prismatic batteries in terms of the energy density. In addition, the pouch battery lowered the entry barrier of lithium ion battery business because of its versatility in the production scale and the ease of manufacturing. In effect, most of small firms targeting niche market adopted pouch lithium ion batteries which Sony had introduced. Like other battery makers, Sony became the victim of field incidents in 2002-2006. Dell's recall of laptop computers using Sony's batteries damaged Sony so hard that Sony seriously considered giving up the battery business. Sony, the first company to commercialize lithium ion batteries, ended up with selling the lithium ion battery business to Murata of Japan in 2016.

Sanyo, a comprehensive battery maker with a worldwide reputation, had led the lithium ion battery industry. Sanyo controlled the flow of technology by the full utilization of battery materials technology. They had the capability to change the anode material of the lithium ion battery. Sanyo's introduction of artificial graphite and natural graphite changed the characteristics of lithium ion batteries. As an example, the lithium ion battery, which had the prominent feature of a long cycle life and a moderate energy capacity, became the high energy capacity battery with a reasonable price by the change of the anode material. The change of the anode material also had influence on the electrolyte and manufacturing methodology significantly. Sanyo's strength in small lithium ion batteries has been successfully extended to the xEV batteries. Automobile makers' trust on Sanyo's capability remains strong, and they think Sanyo is the best source of knowledge in the battery technology.

In the late 1990s Japanese battery makers introduced a new method of battery evaluation technique called the modeling, which provided us with the means to predict the battery performance. Sanyo had led the development of a modeling. Prediction models of the battery performance played an important role in convincing automobile companies to accept lithium ion batteries in spite of potential safety issues. The automobile makers don't like the uncertainty. That's why they use many prediction models in the vehicle design. As regards lithium ion batteries for xEVs, automobile makers wanted to predict the behavior of batteries in the real environment. Sanyo satisfied their needs by providing them with the prediction models. Sanyo's focus on the development of prediction models clearly shows that they take voice of customer (VOC) very seriously for the market-driven research.

Sanyo's knowledge on the prediction model of battery performance had been passed on to Ford when they collaborated with Ford. Volkswagen of Germany also acquired Sanyo's prediction technology through the development project with Sanyo. Automobile makers which had contacted with Sanyo were so impressed by their prediction model that they were beginning to consider the modeling as the tool indispensable to the development of xEV batteries. When Ford developed xEV batteries with Samsung SDI of Korea in 2006-2007, Ford wanted Samsung SDI to use the prediction model, so its knowledge on the battery prediction model was naturally passed on to Samsung SDI by necessity. Afterwards, this modeling tool was conveyed to BMW of Germany when Samsung SDI, more specifically SB LiMotive (SBL), developed xEV batteries for BMW in 2011.

As regards LG Chemical, battery performance prediction model was introduced from Mitsubishi Chemical, another pioneer in the battery modeling, in 2000 when delegation of LG Chemical was invited to Mitsubishi Chemical. However, LG Chemical didn't acknowledge the importance of the battery modeling at that time. LG Chemical, a chemical company, tends to put more focus on the experiment rather than the prediction model.

Sanyo started to develop xEV batteries back in the 1990s as a pioneer in the xEV batteries. They may have thought that the xEV market would give more than enough return on the investment before 2000, but their judgment on the xEV market proved to be wrong. As a result of the premature development activities on the xEV batteries, Sanyo, the leading battery company, got strapped for cash. To make matters worse, Dell's recall of the laptop computers using Sanyo's lithium ion batteries made a dent in the financial status of Sanyo. As a result of financial difficulties, Sanyo was merged to Panasonic in December, 2009. Panasonic was also in serious trouble, but the opportunity Tesla Motors provided to Panasonic galvanized Panasonic after acquisition of Sanyo.

(3) Other battery makers

Asahi Kasai, an inventor of the lithium ion battery, made ATB with Toshiba Battery. Asahi Kasai may have thought that it wouldn't stand a chance in the highly competitive battery market if it moved on its own. Vice President of Toshiba Battery was appointed as the CEO of ATB, and the actual operation of the battery business was conducted by Toshiba Battery because it had a great deal of experience to run the battery business. Strength of Toshiba Battery originated from its balanced battery technology. Thus, ATB was able to compete with the leading battery makers in Japanese lithium ion battery industry in spite of its limited production capacity. In most of the joint companies, expansion of their business through the continuous capital investment rarely occurs. Because of the limitation pertinent to the joint company, ATB could not penetrate into the major market.

GS Yuasa, a lead acid battery maker, joined hands with Mitsubishi Electronics, a mobile phone maker, setting up a joint company called GS-Melcotec. This company produced solely prismatic lithium ion batteries used in mobile phones. Its major market was Mitsubishi Electronics. Pursuant to the market trend of slim mobile phones, competition of prismatic lithium ion batteries used in mobile phones was centered on the development of thin cells ahead of competitors. GS-Melcotec was one of the leaders in this competition. This was made possible by virtue of its strong technical backgrounds in the tool design, which differentiated GS-Melcotec from others in spite of its limited production capacity.

NEC's acquisition of Canada's Moli Energy after the field incidents of Moli Energy's lithium secondary battery gave birth to NEC-Moli. NEC-Moli tried to reduce the cost by adopting low cost spinel manganese oxide (LMO) cathode materials. The LMO interacts with the electrolyte to form a gas, which is the intrinsic problem associated with the spinel structure. NEC-Moli resolved this problem by the surface treatment of the LMO powders. NEC-Moli also developed pouch batteries for the electric vehicle application as a way to make a head start in the emerging market. Hitachi-Maxell was the last participant in the lithium ion battery business in Japan. Hitachi-Maxell acted as a kind of minimum level of entry of the lithium ion battery business. If any company wishes to enter into the lithium ion battery business in the 1990s, it should beat Hitachi Maxell in terms of the market size, production capacity and the battery technology. In this regard, Hitachi-Maxell played an important role in guarding the Japanese lithium ion battery business as a sentry. In actuality,

Hitachi-Maxell became the first target competitor of Samsung SDI and LG Chemical in 2000.

ATB and GS-Melcotec, both of which are joint companies, terminated their lithium ion battery business in the early 2000s when development of batteries one step ahead of others became difficult because of the maturity of lithium ion battery technology. In effect, maturity of lithium ion battery technology was accelerated by the entry of LG Chemical, Samsung SDI of Korea and BYD of China into the lithium ion battery market. NEC and Hitachi group switched their emphasis from small lithium ion batteries to xEV batteries to benefit from early movement in the emerging market. NEC made the joint company called Automotive Energy Supply Corporation (AESC) with Nissan Motor Company, to produce xEV batteries in 2007. AESC makes pouch cells from the electrodes supplied from NEC and assembles the power pack for Nissan Motor's electric vehicles.

Fuji Film, a roll film manufacturer and a newcomer in the battery industry, thought that it needed a sales point to differentiate itself from the leading lithium ion battery makers. Fuji Film, with this in mind, developed the amorphous tin (Sn) anode to replace the conventional graphite anode. It believed that the combination of tin anode with its strong technology background pertinent to coating would give Fuji Film a competitive edge over leading battery manufacturers. However, its expectation turned out to be wrong. The battery market didn't move according to its plan. Mobile phones and laptop computers makers were reluctant to employ Fuji Film's new lithium ion batteries not only because Fuji Film had no track record in the battery business but also because Fuji Film was the only battery maker to employ the amorphous tin anode. The voltage of 3.5V which is lower than that of the other lithium ion batteries by 0.1V was another reason not to employ Fuji Film's lithium ion battery. Fuji Film tried to get it across to the battery industry how good the lithium ion battery incorporated with the tin anode was. However, Fuji Film's multi-pronged effort didn't work out very well. Battery market's apathy towards Fuji Film's new invention forced Fuji Film to give up on the lithium ion battery business.

Battery market is accustomed to the double vendor system. Acceptance of nickel metal hydride batteries in the 1990 in the battery market was based on the fact that two qualified battery makers, Toshiba Battery and Panasonic, provided the batteries that have the same voltage and voltage profile as the conventional nickel cadmium batteries. When Sony, a newcomer in the secondary battery industry, introduced lithium ion batteries in 1991, the battery industry was resistant to accept the novel batteries. Operating voltage was so high and it required the control circuit called the PCM for the safe operation of the battery. The battery industry was beginning to accept lithium ion batteries only when two well-known battery makers, Sanyo and Panasonic, entered the market.

2) Other countries

(1) Chinese battery makers

Entry of BYD of China into lithium ion battery market subsequent to LG Chemical and Samsung SDI of Korea was successful. Its strategy was to rely on manual dexterity of operators to compete with leading companies. BYD decided wisely not to produce cylindrical batteries since cylindrical batteries required a lot of automation processes Japanese battery makers were good at. Instead, BYD focused on prismatic batteries for mobile phones.

BYD's prismatic batteries had a rather peculiar design in that top and bottom parts are closed sideways like an old oriental lunch box. This rather unorthodox design was based on the finding that this configuration facilitated manual assembly of prismatic batteries. As the thickness of the prismatic batteries decreases, the difficulty of inserting the jelly roll into the aluminum can increases, which explains why the yield rate drops in the thin prismatic batteries. The principle behind BYD's cell design was to keep the yield rate stabilized regardless of the thickness of the battery.

Another feature of the BYD process is the coating process, in which it installed many small coating machines instead of a big coater. Although BYD coating process increases the number of operators and makes it difficult to control the uniformity of the cell capacity, it can minimize the damage in case of the mistakes in the electrode making operation. One of Korean battery companies, Samsung SDI, used to form the study group to investigate the BYD process in depth in 2002-2003.

The quality of cells made by BYD depends upon the skills of operators, so wide spread of the battery quality was inevitable. BYD accepted this intrinsic limitation and came up with its own marketing strategy. BYD's marketing methodology was to match cells with different quality level with the market of specific requirements. In so doing, it could sell almost every cell regardless of the quality. The manufacturing methodology and the quality control method enabled BYD to sell the prismatic lithium ion batteries at a very low price. The fact of the matter is cost was the most prominent advantage of BYD.

BYD was very aggressive in the investment of the battery plant and expansion to the pertinent areas. Their aggressiveness in the battery investment paid off in electric vehicle batteries. BYD had also grown to become an electric vehicle maker, leading a Chinese electric vehicle market. In the beginning, BYD used lithium iron phosphate (LFP) cathode in the electric vehicle batteries to meet the safety requirement of Chinese vehicles. BYD is keen to add the NCM batteries to the product portfolio for the performance improvement and, more importantly, to compete with Japanese and Korean battery makers in the international market.

Winston, previously known as Thundersky, came up with a method to make a large lithium ion battery at the lowest cost in the early 2000s, which is currently called plastic can lithium ion batteries. They used LCO and LFP as the cathode materials because high performance nickel base cathode materials were unavailable to Chinese battery makers at that time. Winston's batteries were used in the electric bus of China. Although some incidents like fire of the electric bus took place when LCO batteries were used, replacement of LCO batteries with LFP batteries resolved the safety issue.

International Battery (IB) of the USA was a power pack maker which supplied power packs to the US Army using Winston's large plastic can LFP batteries. However, serious energy capacity deviation of cells supplied from Winston was troublesome. In order to resolve the quality issues, IB decided to make its own cells using Winston's technology and, with this in mind, sent the engineers to Winston and made them stay there for one year for the acquisition and comprehensive documentation of plastic can battery technology. After the acquisition of the technology associated with large format lithium ion batteries, IB constructed the battery plant in Philadelphia in 2008. Unlike ordinary lithium ion battery plants, it had no dry room and, more surprisingly, the cells didn't go through the aging process for the screening of defective cells, so the plant looked more like a lead acid battery plant rather than a lithium ion battery plant in terms of the configuration of the plant.

When IB was building a battery plant in the USA, it was contacted by Hyosung of Korea for the possible collaboration. Hyosung's chairman was appointed as the Chairman of Federation of Korean Industries in 2007. Capitalizing on this opportunity, Hyosung wished to expand into the field of large lithium ion batteries used in electric vehicles. Hyosung had conducted the lithium ion battery development in the 1990s until it terminated all the battery activities in the Korean financial crisis in 1998. Thus, Hyosung was not completely ignorant of the battery business. Hyosung's delegation made a visit to the IB's battery plant for a couple of times, and both parties agreed to build the battery plant jointly in Korea. However, Hyosung couldn't find any suitable market in Korea. The market Hyosung aimed at was the electric vehicle market, the highly competitive and challenging market, but the market which IB's large plastic can lithium ion batteries can enter was the UPS and golf cart at best. The time had come for Hyosung to fish or cut bait. After unfruitful efforts to find the suitable opportunity in the area of lithium ion battery business, Hyosung decided to drop all the activities associated with lithium ion batteries in 2010. Instead, Hyosung switched its efforts to the development of ESS system in which it has strong technology foundation through the transformer business. As a result, Hyosung was able to have a large role in the ESS industry.

(2) Battery makers in North America

Eveready of the USA belongs to a leading group in the primary battery industry. Eveready built a lithium ion battery plant with the largest dry room in 1997 with a full of confidence on the lithium ion battery technology. However, against its expectation, it struggled heavily to obtain a decent production yield. Six months after the opening ceremony, Eveready decided to shut down the lithium ion battery plant. It was a big blow to the ego of the US battery industry. Prime time of the US battery industry seemed to pass in the age of material dominant battery age, in which data generation and analysis become more important than creativity. After Eveready failed to make lithium ion battery technology working commercially, senior managers of Eveready took a drastic measure. They decided to give up on all the secondary battery business. They might have thought that their focus on the main business, primary batteries, would reinforce their capabilities in the battery market. Eveready sold the nickel metal hydride business to Moltech located in Tucson, Arizona.

Moltech was a R&D company which was developing lithium-sulfur (Li-S) batteries. Moltech had discussed the technology transfer of lithium-sulfur batteries with LG Chemical. However, Moltech was not able to sign the agreement with LG Chemical of Korea in spite of the discussion on the terms and conditions of the agreement on and off for two years. Somebody had to take responsibility for the lost time and excessive expenditure accruing from the discussion with LG Chemical. BOD of Moltech asked the CEO and the founder of Moltech to step down. New CEO of Moltech who got promoted from within acquired the nickel metal hydride battery business from Eveready and established the battery business foundation upon which Moltech could develop lithium-sulfur batteries in the long term. Moltech's acquisition of nickel metal hydride business from Eveready was a big surprise. Some people in the battery industry said that a small fish swallowed a whale.

From the mid 2000s, the US government supported the battery industry in the USA financially. Cash-rich battery firms were very active in the acquisition of lithium ion battery technology. The best source of knowledge in lithium ion batteries was Japanese battery makers, Samsung SDI and LG Chemical of Korea, but they did not agree to transfer their technology to the US firms for the obvious competitive reasons. As an alternative, the US firms approached other Korean battery companies and reached the agreement for the technology transfer. KoKam sold its battery technology to Dow Chemical, Enerland to A-123 and Enertek to Enerone. Three Korean battery makers have one thing in common. They all made pouch type lithium ion batteries for the niche market. The potential problem confronting the US battery companies was that these Korean battery makers which provided the battery technology to the US companies didn't have enough experience in the mass production of lithium ion cells.

Dow Chemical and Enerone were rather conservative in proceeding the battery business, but A-123 was different. After merging of Enerland of Korea, A-123 constructed the mass production plant in a large scale for the entry of electric vehicle market. However, its aggressive investment strategy didn't pay off. On the contrary, its aggressiveness in the investment and expansion of the battery business led to the premature demise. Its target batteries were pouch batteries incorporated with lithium iron phosphate (LFP) cathode materials. Performance and cost of lithium iron phosphate batteries failed to exhibit competitive edge over lithium ion batteries incorporated with NCM cathode. In addition, A-123 opted for the high cost design in terms of the battery materials and manufacturing methodology for the expeditious entry of the market, which aggravated the bottom line of the company seriously. In spite of the efforts to capture the market with the strong financial support of the US government, A-123 failed to acquire the competitiveness. After several years' struggle, A-123 was sold to Wanxiang of China. A-123 had become the second victim of the US lithium ion battery business after Eveready. A-123's failure story gave us the opportunity to figure out what went wrong in the attempt to pursue the lithium ion battery business in spite of all the financial support of the US government.

3) The road to success

There are two ways in front of belated starters: the road to success and the road to failure. Nobody wants to take the road to failure, but in actuality most of late starters take the road to failure in spite of themselves. In order for the late starters to succeed in the lithium ion battery business, strange as it may sound, it is imperative that they restrain their creative mind. This explains why the US battery makers could not make the success story in the lithium ion battery business. The lithium ion battery is different from the conventional batteries in that it is a material-oriented battery, so data generation and the analysis of the accumulated data are more important than the creative idea. In effect, data handling activity is the essence of lithium ion battery development. Therefore, if you want to start the lithium ion battery business, the first thing you have to do is to hire lithium ion battery engineers who not only have the intimate knowledge on the lithium ion battery business but have sufficient mass production experience. Based on their experience and knowledge, you install the equipment which is same as that of the leading companies and make cells which have similar characteristics to the leading companies. As time goes by, you get familiar with the manufacturing process of the lithium ion batteries. This is when you jump to a new stage by taking advantage of differentiation points and creative ideas. This is what successful late starters, Samsung SDI and LG Chemical, had done.

As regards the manufacturing methodology, both Samsung SDI and LG Chemical took nothing for granted, so they didn't change any process before they had full understanding of the Japanese manufacturing method. Their conservative approach not to rush into introducing a new method in their process really paid off and made a big difference. In contrast, late starters who failed did quite opposite, which is called the road to failure.

The heavier you are, the harder you fall. In the late 2000s, A-123 of the USA was a big guy. That's why it fell so hard. A-123 hired many battery engineers mainly from Korea, but most of them were young engineers of less than five years experience. In addition, among the senior management group and BOD members, there was no one who had the mass production experience in lithium ion batteries. Most of BOD members including the CEO had the background in the electronic industry, so they considered lithium ion battery business as the assembly business like a laptop computer business. They chose the LFP batteries as the target batteries just because one of the founders of A-123 had a great deal of experience in the development of LFP cathodes without seriously checking out the pros and cons associated with the LFP battery. The advantage of LFP batteries is said to be the high safety feature and low cost. Safety test shows that it is much safer than NCM batteries, especially in the nail penetration test. Battery engineers prefer high safety batteries because they can reduce the cost by simplifying the pack design. In other words, advantages of the high safety battery should be reflected in the cost. Otherwise, high safety battery is meaningless business-wise. LFP batteries failed to reduce the pack cost. Rather, A-123 had to use the same pack design as the NCM batteries. As regarding the cost of LFP cathode, it did not exhibit any sign of cost advantage compared to the NCM cathode possibly because of the low yield rate and the patent barrier.

In the perspective of manufacturing process, it was a really challenging task to make a sound electrode using LFP cathode materials because of the adhesion problem between electrodes and aluminum foil, which not only lowered the yield rate but made it hard to maintain the uniformity of the energy capacity. The most important thing in the battery business, especially in the electric vehicle battery business, is to get the priority straight. First, we have to design the battery chemistry that can enable us to achieve the cost target and then go ahead with the improvement in the cell performance. This is the whole idea of the chemistry freezing. After chemistry freezing, we can't change the cell chemistry and cell design in principle. Change of the cell design and the cell chemistry after chemistry freezing would delay the development time significantly. Because of this rigidity in the development process, it is necessitated to design the cell with the cost-effective materials and process in the first place. However, A-123 was too hasty to follow this basic rule. It tried to achieve the cell performance target without regard to the cost. Unlike other battery makers, it used a very expensive material, the example of which is the electron conducting material called vacuum growth fiber (VGF). It also adopted the abnormally high n/p ratio to extend the cycle life, which obviously increased the cost because of the redundant anode materials. It also made the aging period much longer than other battery companies to acquire a long cycle life. As a result of all these actions, A-123's cell design was no longer competitive due to the use of expensive materials and the production cost. From the business perspective, A-123's battery business was doomed before it started. The more it sells, the bigger the deficit becomes because of the problematic cost structure.

With regard to the lithium iron phosphate (LFP) cathode, there is another story relative to this cathode material. Hanwha Chemical is one of the major petrochemical firms in Korea. In order to reduce its dependence on the pertrochemical business, Hanwha Chemical expanded into the field of solar cells. In the development of the pertrochemical products, Hanwha Chemical's engineers acquired a great deal of technological know-hows associated with supercritical hydrothermal synthesis (SHS). As an application to the SHS technology, it decided to enter into the battery materials business by developing lithium iron phosphate (LFP) cathode materials. Confident that it can make a better cathode materials than the NCM cathode makers adopting co-precipitation process, senior managers of Hanwha Chemical didn't bother to hire battery engineers who have the experience in the battery and battery materials. Furthermore, Hanwha Chemical's engineers didn't pay attention to the information on the equipment that the leading companies were using. Instead, they designed the process consisting of SHS and spray drying process, both of which had not been adopted in the battery material business because of the difficulties in controlling the powder size and morphology. Hanwha Chemical's manufacturing process was more expensive than the conventional LFP process using solid-solid reaction or hydrothermal process because excessive lithium was consumed. However, Hanwha Chemical's engineers thought that cost wouldn't be a problem because their LFP exhibited better performance. They failed to understand how competitive the battery business was in terms of the cost.

After completion of the development, Hanwha Chemical built the LFP plant and supplied LFP cathode materials to BYD and Wanxiang of China. Several months after the inception of the supply of LFP, Wanxiang and BYD notified Hanwha Chemical of the quality issues and asked to ship back LFP cathode materials because the LFP cathode materials failed to fit their requirements. Hanwha Chemical tried to rectify the problems pertinent to the LFP cathode materials for a couple of years, but every effort was proven to be ineffective. Three hundred tons of malicious inventory worsened the situation. Finally, Hanwha Chemical terminated the LFP business in December, 2013 because the cost structure wouldn't allow Hanwha Chemical to make a profit even though it resolved the quality issues.

The road to success is also strongly related to the development of a good battery. Depending upon the experience of each person, battery engineers in the battery industry will define a good battery differently. The

1990s was the decade of portable electronics. Laptop computers, cellular phones and the camcorder, which were called the 3C applications, had led the battery market. In the battery for portable electronics, competition boiled down to who would make smaller, thinner and lighter batteries because the sales point of portable electronics was the weight, volume and shape rather than the function. Therefore, competition to make a high energy density battery was very tight. Energy capacity improvement called the version-up took place every ten months. Any battery company that was left behind in the competition of high energy density battery development had lost the market. As a result of this tight competition to develop high energy capacity batteries, energy capacity of lithium ion batteries doubled for the period of ten years. Therefore, in the 1990s the high energy capacity battery was a good battery because it gave us the competitive edge over competitors in the portable electronics market.

In leading battery firms, the energy capacity was increased by developing a new core technology, and therefore, the balance was maintained in the cell performance regardless of the energy capacity. In contrast, battery companies which lacked knowledge and experience had to increase the energy capacity at the sacrifice of safety margin to compete with the leading companies. As a result, an awkward looking batteries in terms of the balance of cell performance were created. They just increased the energy capacity by merely densifying the electrodes without any additional new technology. Unreasonable demands of high capacity batteries from portable electronics makers forced battery makers to make many unreasonable attempts. The cut-throat competition to make high energy capacity batteries ahead of others in the 1990s resulted in field incidents like fire or explosion in 2002-2006. In the first half of 2000s, practically every lithium ion battery makers in Japan and Korea got victimized by field incidents. As a remedy to a fire risk battery, battery makers were keen to reinforce the safety features by increasing the design margin through the introduction of new technology. Ceramic coating on the separator is one example to make lithium ion batteries safer. In summary, the 2000s was a safety-conscious decade, so a safe battery was considered as a good battery.

As described above, the definition of a good battery changed with time and the market trend. Nevertheless, there are some features associated with a good battery regardless of the time and the market trend. First, the battery with a well-balanced cell performance is a good battery. Battery is like a round balloon. A good battery is the one that maintains the round shape under any circumstances. In order to do that, energy capacity, rate performance, temperature performance and cycle life should be balanced. Second, the battery with a high cycle life is a good battery. Strictly speaking, cycle life is not the cell performance. It is the representation of the cell degradation. High cycle life battery is like people who maintain their youth and virility for a long period of time. Moreover, one of the prominent features of the battery made by the leading companies is a long cycle life. Therefore, cycle life can be arguably construed as the criterion of a good battery. Third, the battery with less energy capacity deviation is considered as a good battery. Uniformity of the energy capacity of cells makes the design of power packs incredibly simple. Combination of the uniformity of energy capacity with a long cycle life gives birth to a battery with a good state of health (SOH). Thus, the most reasonable way to rank the battery firms in terms of the technology level is to evaluate the uniformity of cells they produce.

4) Technology flow

Lithium ion batteries were created in Japan in 1991 and Japanese battery companies had dominated the world market for ten years in the 1990s. In this regard, we don't hesitate to call Japanese major makers, Sony, Sanyo and Panasonic, original battery makers. They had the power to change the battery market whenever they liked, which is called the merit of the original makers. They had been ahead of the curve since 1994 until Sanyo was merged to Panasonic in December, 2009.

There was a flow from Japan to Korea in 1996 as a result of Japan's decision to lift the restriction on the outflow of lithium ion battery technology and the conflict between battery engineers and senior managers in Sony. The technology transferred at that time was basically the Sony's lithium ion battery technology, which was somewhat different from Sanyo's battery technology. Sanyo's technology was more materials-oriented than Sony's battery technology. In fact, Samsung SDI preferred to acquire Sanyo's lithium ion battery technology, but its technology was hard to come by.

Panasonic Electric Vehicle Energy (PEVE) is a joint company of Toyoto, an automobile company, and Panasonic, a battery company. In the beginning, there was a balance between Toyota and Panasonic, but the equilibrium state didn't last long. Toyota had a much larger say and role by the increase of the stake of PEVE, which made some Panasonic engineers with a liberal mind within PEVE leave their company. These Panasonic engineers came to Samsung SDI in 2004-2005, and formed the foundation for the Samsung SDI's xEV battery technology. Because of the technology inflow from Panasonic, Samsung SDI's prismatic battery for xEVs is similar to that of Panasonic in terms of the basic design and the chemistry.

After the elapse of five years since the inception of the battery business, battery engineers in Samsung SDI and LG Chemical acquired the capability to develop the battery for themselves. Some of them were beginning to leave their companies in the mid 2000s for the better career opportunities. Among them junior members went to the USA to join A-123, and senior members went to China for the consulting job. In addition to major battery firms, there are small battery firms, Kokam, Enerland and Enertek, that make pouch batteries for the niche market. Their pouch battery technology was transferred to Dow Chemical, A-123 and Enerone in 2008-2009.

We can conveniently classify Chinese battery makers into three groups. The first group is the battery makers that developed lithium ion battery without any assistance from Japan or Korea. BYD belongs to the first group. Plastic can prismatic battery makers such as Winston, Calb, Sinopoly and GBS also developed batteries on their own. The second group is the battery makers that have been influenced by Japanese technology. ATL is the typical example of the second group. ATL of China was founded by Chinese engineers educated in the USA in 1999. TDK of Japan had established the connection with ATL by equity investment, and the ownership of ATL was transferred to TDK of Japan in 2005 by TDK's acquisition of 100% stake of ATL. ATL has been competing with Samsung SDI in Samsung Electronic's smart phone market. ATL's pouch batteries have been successful in making inroads into Samsung SDI's prismatic batteries in the smart phone market. In addition, the ATL's pouch battery exhibited its superiority in comparison with Samsung SDI's pouch battery in Samsung Electronic's Galaxy Note 7 in 2016 in spite of the field incidents in the replaced phones. ATL also competes with Samsung SDI in the BMW market. ATL supplies aluminum can prismatic batteries to BMW of Germany. The third group is the battery makers with the technology inflow from Korea. Lishen, Wanxiang and BAK which have been active in hiring Korean battery engineers belong to the third group.

Johnson Control, an automotive parts firm in the USA, acquired the lithium ion battery technology from SAFT, and developed its own battery technology based on the SAFT's battery technology, so Johnson Control's lithium ion battery technology has the stand-alone characteristics. Johnson Control hired some Korean battery engineers who used to work with A-123 after A-123 was sold to Wanxiang. Johnson Control has maintained a good working relationship with Ford.

There was a strong technology flow from Korea to Germany. Unlike the USA, Germany wanted to acquire lithium ion battery technology that had been verified in the market. Global automotive parts firms came forward to carry out this mission. In the beginning, they approached the Japanese battery makers, but Japanese battery makers were reluctant to transfer their battery technology to German automotive parts firms. Next target was major battery makers in Korea. At that time, LG Chemical collaborated with GM in the area of PHEV (plug-in hybrid electric vehicle). There was no room for the German makers.

Samsung SDI worked together with Ford for the development of HEV batteries, but there were some senior managers within Samsung SDI who preferred European companies to Ford simply because they were more familiar with European companies. Samsung SDI had braun tube plant in Europe. Aware of the conflict within Samsung SDI in association with a development partner, Bosch approached Samsung SDI for the acquisition of xEV battery technology and managed to acquire xEV battery technology from Samsung SDI through the joint company called SB LiMotive (SBL). Continental, a German automotive parts firm, followed suit by making a joint venture with SK innovation. Germany is the mecca of the premium brand cars. German automotive parts firms have good battery technology as well as the marketing capability, so they may emerge as the formidable players in the xEV battery market in a perceivable future.

In Japan, there was a technology flow from lithium ion battery firms to the automobile companies. Panasonic provided battery technology to Toyota, forming PEVE. GS Yuasa collaborated with Mitsubishi Motors to form Lithium Energy Japan (LEJ) in 2007 and made another joint company called Blue Energy with Honda in 2009. NEC, which produces lithium ion pouch batteries, furnished the battery assembly technology to Nissan Motor Corporation, creating Automotive Energy Supply Corporation (AESC). Although history is too short to make any judgment, marriage between the automobile firms and the battery companies seems to be successful.

Korean battery makers seem to prefer automotive parts firms to automobile companies. LG Chemical supplied lithium ion battery power pack technology to Hyundai Mobis, an automotive parts company within Hyundai Motor group, and made a company called HL Green Power in 2010, which is a 51:49 joint company of Hyundai Mobis and LG Chemical. SB LiMotive (SBL), which was a 50:50 joint company of Bosch and Samsung SDI, was founded in June, 2008. Samsung SDI furnished lithium ion battery technology to Bosch, and in return for this, Bosch took responsibility for capturing the German market. SK innovation also joined with Continental of Germany by providing lithium ion pouch battery technology to Continental, giving birth to a joint company called SK Continental E-motion. Hyundai Mobis, Bosch and Continental are automotive parts companies which provide automotive parts to automobile companies. Marriage between lithium ion battery makers and automotive parts firms didn't last long. Samsung SDI and Bosch got separated in September, 2012. Following in the footstep of Samsung SDI, SK innovation terminated its association with Continental in 2013.

One of the reasons for Korean battery companies to prefer to collaborate with automotive parts companies rather than automobile makers is that somewhere along the line they want to enter into the electric vehicle business like BYD of China and Tesla Motors of the USA. The EV business is somewhat different from the conventional automobile business and, more importantly, batteries are the key component to determine the competitiveness. In this regard, the sizable battery companies may stand a good chance in the future automobile market. Based on this reasoning, Korean battery companies may have tried to form the vertical integration through the collaboration with automotive parts companies. However, the problem is that automotive parts firms are too gigantic for the battery makers to wrestle with. LG Electronics started automotive parts business and Samsung Electronics is keen to prepare for the electric vehicle business with the focus placed on a self-driving car. Samsung's equity investment in BYD and the in-depth business discussion with Fiat Chrysler Automobiles (FCA) on the possible purchase of Magneti Marelli reflect Samsung's desire to enter into the electric vehicle business.

After Volkswagen's emission scandal in September, 2015, Volkswagen seems to have changed its view on electric vehicles and a self-driving car. Chairman of Volkswagen announced "Together-Strategy 2025" in 2016. Volkswagen's senior managers proclaimed that they would grow the battery technology as the core competence of the Volkswagen group and with this in mind Volkswagen has been discussing with major battery makers in Korea and Japan for the possible merger and a joint venture.

Chinese government's subsidy ban on the NCM batteries in the electric bus thwarted Korean battery makers' plan to expand into the field of Chinese electric bus market. Samsung SDI and LG Chemical completed the construction of the battery plant in China in 2015 for the preparation of the Chinese electric bus market. The Chinese safety standard is very high because it is based on the safety features of the LFP battery which is known as the safest battery among lithium ion batteries. After several years' efforts, the NCM battery of LG Chemical and Samsung SDI passed the Chinese safety test in 2012, so they constructed the battery plants in China in the hope that the plants in China would facilitate the operation in the Chinese electric bus market. However, unexpected subsidy ban on the NCM batteries threw cold water on their plan. It is not hard to figure out why Chinese government enforced a subsidy ban on NCM batteries.

First, Chinese government may have thought that the inflow of technology from Korea was too fast, so it wanted to put a restriction on the inflow from Korea. Second, because of the bad memory associated with fire or explosion of the electric bus powered by LCO batteries, Chinese government may have believed that simply passing the safety test would not guarantee the absolute safety of the battery in a real environment. To make matters worse, explosion of Samsung Electronics' Galaxy Note 7 powered by Samsung SDI's pouch batteries, in August, 2016 may reinforce Chinese government's belief that NCM batteries are not safe in the electric bus although LCO batteries were used in Galaxy Note 7. From the perspective of Samsung SDI and LG Chemical, the most realistic way to resolve this issue is to supply LFP batteries to Chinese electric bus market concurrently with the continuous persuasion of the Chinese government.

There are over 1,000 battery companies in China, so the need naturally arises to determine which companies are eligible for the subsidy. Based on the intent of reorganizing the battery industry in China, Chinese government stipulated that battery makers are to be registered as subsidy beneficiaries to receive the subsidy from Chinese government regardless of the nationality. In order to be registered as subsidy beneficiaries, the battery company should have over one hundred research engineers and the production capacity of over 200 MWh which is equivalent to 10,000 electric vehicles. This regulation is considered as the another entry barrier for LG Chemical and Samsung SDI. Other companies like SK innovation and Panasonic are out of the race because they don't have the battery plant in China. LG Chemical and Samsung SDI failed to be registered twice in 2016 probably because their plant operating period was too short. Therefore, chances are Samsung SDI and LG Chemical will be likely to be registered as subsidy beneficiaries before the end of 2016 or early 2017.

Inflow from Korea is necessitated to grow Chinese electric vehicle market. Without the technology inflow from Korea, Chinese electric vehicle industry would suffer from the lack of battery technology, especially the mass production technology of batteries. The common weakness of the Chinese battery industry is the lack of mass production technology of lithium ion battery. For example, no Chinese battery companies had experienced field incidents of laptop computers in 2002-2006. Experience of field incidents is an important part of mass production technology of lithium ion batteries as long as the damage of the filed incidents is not fatal.

PART

3

KOREAN BATTERY INDUSTRY

09

:

Battery makers

1) The major companies

LG Chemical and Samsung SDI became the first lithium ion battery makers outside Japan in 1999-2000. They had competed with each other in the portable electronics market and extended their competition into the area of xEVs. Their internal markets, Samsung Electronics and LG Electronics which produce mobile phones and laptop computers, were big enough to provide the stepping stone to their lithium ion battery business, especially in the initial stage of the battery business. The competition between Samsung SDI and LG Chemical was diversified by the involvement of SK innovation. SK group announced its intention to enter the lithium ion battery market by focusing on xEV batteries in December, 2007. SK group judgmatically switched the target market from portable electronics to xEVs because unlike Samsung and LG, SK group does not have the internal market of batteries.

LG Chemical, Samsung SDI and SK innovation form the leading group. The reason for including SK innovation in the major makers in spite of the belated start is because SK innovation has what it takes to be a leading battery maker. Unlike semiconductor and display business, investment capability of the production plant is not the entry barrier. In fact, any small company can build the lithium ion battery plant because the investment cost is small compared to other businesses. The entry barrier of lithium ion battery business is the R&D capability, which is to conduct the battery development by hiring top-notch scientists and engineers continuously and competitively. Only a few companies have this level of R&D capability. Another criterion is the commitment of the chairman of the company to the battery business. In contrast with other businesses, lithium ion battery business has many ups and downs and, more importantly, ups and downs take place quite frequently. Decline in the rate of operation due to the oversupply and the event of ship-back or recall caused by the field incidents make battery companies wobble severely. It is only the steadfast commitment of the chairman to lithium ion battery business that can enable battery makers to withstand the hardship and continue to proceed the battery business regardless. As a chemical company, SK innovation outranks its competitors in R&D in terms of R&D force, infrastructure and output. As regards the leadership of the chairman, chairman and vice chairman of SK group have been leading the xEV battery business and their commitment to the xEV battery business remains strong and steadfast.

In conclusion, Samsung SDI, LG Chemical and SK innovation became the top three makers of Korean battery industry because they fill the bill to be included in the leading group in terms of R&D capability and chairman's commitment to the battery business.

(1) LG Chemical

Based on the policy to focus on the next generation battery, LG started the battery development activity through the development of LPB with AEA of the UK. After LG gained an understanding of the battery business, it moved all the battery activities including a LPB project from LG Metals to LG Chemical, the main company within LG group, for the aggressive investment in the battery business in April, 1996. Upon receipt of the battery business from LG Metals, LG Chemical changed the strategy from its focus on the development of the next generation battery to the foundation of the battery business base through the commercialization of lithium ion batteries. LG Chemical's intention was to establish the battery business base and the pertinent infrastructure in parallel with the pursuit of the next generation battery.

LG Chemical wanted a battery expert with a reputation in the battery industry to lead the battery business. Arthur D. Little (ADL) of the USA had been developing lithium ion polymer batteries in 1990s. LG Chemical interviewed the team leader of the battery development team of ADL. Arcotronics, an Italian battery equipment maker, hired a battery expert to provide lithium ion battery technology along with the battery equipment. Arcotronic's battery expert who used to work with Moli Energy of Canada was interviewed by LG Chemical for the same position. Incidentally, both of them were Canadians. The cold country like Canada is strong in the battery technology because the cold weather requires electric coating on the metal structure. That's why countries like Canada are traditionally strong in electrochemistry. Two Canadian battery engineers wanted to work with LG Chemical, but the problem was the poor infrastructure of Korea in the 1990s associated with the education system of children from abroad. Their association with LG Chemical ended there.

There was a discrepancy of opinion between senior managers and the battery engineers regarding lithium ion battery business strategy in Sony. Sony's battery engineers wanted to grow the battery business in competition with Sanyo and Panasonic. However, Sony's senior managers considered the battery business as the parts supplier for their main business. Whenever there is an argument between senior managers and the working group, senior managers usually win. That's what had happened in Sony. Senior managers of Sony made many battery engineers leave Sony in the early 1996. Up until early 1996, acquisition of the battery equipment and battery materials was a serious problem for battery companies outside Japan. Japanese component suppliers and equipment makers were resistant to sell their products to the foreign battery companies because it might hurt their relationship with Japanese battery makers. These restrictions were also lifted in 1996. Japanese battery companies allowed material and equipment suppliers to sell their products to foreign firms without any restriction. In conclusion, Japan decided to reveal lithium ion battery technology in 1996 in the hope of expanding the lithium ion battery market.

Acknowledging the change in the Japanese battery industry, LG Chemical had switched its attention from Canada to Japan, and hired a research engineer who used to work with Sony, and let him lead the development of lithium ion batteries in 1996. LG Chemical constructed the pilot plant for 18650 cylindrical batteries and sold the batteries to the after-market of Hong Kong, Taiwan and China to understand the battery business before the mass production is initiated. Based on the confidence collected from the sales of cylindrical cells, LG Chemical constructed the battery plant and started mass production in 1999.

Historically, LG group, previously known as Lucky-Goldstar group, is the combination of two separate firms, Goldstar and Lucky. Goldstar, currently known as LG Electronics, is the oldest electronic company in Korea. Lucky, currently identified as LG Chemical, is a chemical company that originally produced toothpaste, soap and cosmetics. Lucky-Goldstar group changed its name to LG in the early 1990s to strengthen its solidarity. As LG became larger, it was divided into three conglomerates: LG, GS and LS. Therefore, LG, GS and LS group have the same root and consequently similar business culture. LG considers it very important to maintain the continuity of technology. LG wants its tradition relative to the technology to flow smoothly like a water flow in the river. Therefore, when the need arises to change the management of the business, LG is very careful not to discontinue the flow of technology by appointing people who have the intimate knowledge on the pertinent technology. LG tends to think that technology is something that should be cultivated like a tree and a flower. In this respect, LG has the mind of a farmer.

LG is famous for making an investment with extreme caution and thoughtful consideration. In order to get an approval for the investment, the pertinent department has to go through many steps of review. In addition, after the investment senior managers check out the progress on a regular basis to determine if there is any oversight in the investment decision. However, LG Chemical's investment on the plants for xEV batteries is recorded as a fly in the ointment. People in the battery industry couldn't understand why a cautious LG Chemical made an investment in the uncertainty so aggressively. In the midst of LG Chemical's xEV battery investment, many people in the battery industry, including LG chemical's battery business, noticed that the market was sending a signal of slowing down. LG Chemical should have stopped the investment at that moment, but to the surprise of the battery industry, LG Chemical sticked to the initial investment plan. As a result of excessively aggressive investment on the xEV battery plants, xEV battery industry was forced to be rushed into the commercial stage unprepared.

(2) Samsung SDI

Samsung started the battery business activity at Samsung Electronics, the main company within Samsung group. Samsung Electronics wanted to acquire hands-on experience in the battery business through the development of mass production technology of commercially verified batteries rather than developing the next generation battery. Samsung Electronics acquired the nickel metal hydride battery technology from Ovonics of the USA to apply these batteries to the cordless phones. Samsung Electronics constructed the pilot plant for nickel metal hydride batteries for the development of mass production technology, but battery engineers couldn't make a decent cell because of the breakage of electrodes. After the period of unfruitful efforts and incessant trial and error, Samsung group finally came to the conclusion that Samsung Electronics might not be suitable for the battery business, so Chairman of Samsung group moved the battery business from Samsung Electronics to Samsung SDI for the consistent proceeding of the battery business in 1995. Samsung SDI was the display firm which made the braun tube. TFT-LCD was developed at Samsung SDI and was transferred to Samsung Electronics afterwards and Samsung SDI was developing plasma display panel (PDP) at that time.

Samsung SDI acknowledged that Japan decided to disclose the lithium ion battery technology by releasing the battery engineers and allowing the materials and equipment makers to sell their products to foreign companies in 1996. Samsung SDI wanted to make a head start by capitalizing on this opportunity, so it hired a wide spectrum of battery engineers ranging from cell development, manufacturing, quality control, pack development and equipment design from Japan, and made them head up the departments of the Samsung SDI's battery business division. Samsung SDI started the mass production of lithium ion batteries successfully in 2000. Encouraged by the progress of mass production of batteries, Samsung SDI planned to change the competition landscape at a brush by introducing new battery materials. Battery engineers of Samsung SDI developed nickel-base cathode materials and applied new cathode materials to their cells to outperform the LCO batteries produced by Japanese makers, but their try didn't work out very well in the scale-up stage. Coating slurry got agglomerated as a gell, so they couldn't even make the electrodes, much less the batteries.

The higher the expectation is, the bigger the disappointment is. Disappointed at the outcome of the attempt to make a quantum jump, Samsung SDI's senior managers took a drastic measure in 2002. They decided to drop all the activities associated with battery materials development and switch their efforts from battery materials development to the search for the low cost battery materials. Pursuant to Samsung SDI's decision to focus on searching for the battery materials, Cheil Industries, subsidiary of Samsung group and a development partner of Samsung SDI, decided to give up their electrolyte business and sold the battery-related materials business including NCM cathodes to Ecopro located close to LG Chemical's battery plant. Ecopro had completed NCM cathodes development with the help of LG Chemical and supplied cathode materials to LG Chemical. Ironically, NCM cathode materials, the development of which was initiated by Samsung SDI, was completed by LG Chemical. Cheil Industries' material division was merged to Samsung SDI in 2014 to rekindle its interest in battery materials, twelve years after discontinuing battery materials development activity.

Thanks to the Samsung SDI's strategic decision regarding the development of battery materials, Samsung SDI was able to move fast because the battery engineers didn't have to develop their own battery materials. However, senior managers' decision not to develop battery materials was beginning to have adverse effect on the xEV batteries.

Major battery makers got victimized by the field incidents in 2002-2006. Sony, Sanyo, Panasonic and LG Chemical suffered from the aftermath of the recall of laptop computers powered by their lithium ion batteries in the second half of the 2000s. However, HP's decision not to recall laptop computers using Samsung SDI's 18650 cylindrical batteries in spite of severe field incidents for six months in 2003 saved Samsung SDI from being demolished by the field incidents. Samsung SDI became a lone survivor. Samsung SDI knew how to capitalize on the misfortunes of competitors. Samsung SDI put all the available resources to penetrate into the markets which their competitors used to serve before the recall incidents. In the end, Samsung SDI became a top maker in 2011 in terms of the sales volume of small lithium ion batteries.

Samsung and LG have been in rivalry for a long time. Longevity of their rivalry period stems from the fact that both of them have so many differences and similarities in many aspects. While LG has the mind of a farmer, Samsung has the characteristics of a nomad. As for Samsung, technology is something that should be acquired and managed. In fact, technology management is the core competence of Samsung. When Samsung changes the management, it discontinues the flow deliberately and make the new direction for the flow. It is like stairs as compared with water flow of LG. Samsung's management style is very effective in reforming the organization drastically.

In LG, individual performance is the driving force for the movement of the business. In contrast, system drives the business in Samsung. Design review (DR), completely independent quality control system and the group value engineering (GVE) are typical examples of Samsung's system. The GVE is the activity to cut costs by simplifying the electronic circuits. In batteries, the GVE is the activity to search for low cost materials. The philosophy behind Samsung's operating system is that an individual can make a mistake, but the system doesn't. Of course, there is some truth in that, and in most cases, Samsung's system worked out very well. However, somewhere along the line we may need more than following the system.

Future market trend may be different from what it is now. It could be the individual performance and creative mind that determine which company will be the winner in the xEV battery business. Therefore, Samsung SDI should ask itself the following question. "Can Samsung SDI and Samsung group possibly cope with the future challenge of lithium ion battery business for xEVs with the system alone?"

Samsung believes that when it comes to lithium ion battery technology, Japan is the best source of knowledge because of the merit of the original makers. Based on this reasoning, Samsung SDI has been hiring accomplished professionals from Japan since 1996. Unlike western people, Japanese people have a tendency not to speak out. On the other hand, Korean scientists and engineers tend to find the solution on their own rather than asking questions. In other words, Japanese battery engineers are bad teachers and Korean battery engineers are lousy students, so they don't mingle with each other in the same organization. In actuality, it is one of the challenging tasks of Samsung SDI to make a homogeneous working group regardless of nationality. Discord within the organization due to the diversity of people is the problem Samsung SDI has to wrestle with to maintain the competitiveness for years to come. Nevertheless, strength of Samsung SDI and Samsung group comes from the diversity of people. Samsung's willingness to recruit qualified and accomplished professionals regardless of nationality is really commendable in a Korean society where homogeneity is preferred to diversity.

(3) SK innovation

SK's battery development history is as long as that of Samsung and LG. SK started the lithium ion battery development activities back in the 1990s at SKC whose core business is the video tape production. Thanks to the SKC's excellent coating technology acquired through video tape business, battery engineers in SK were able to develop lithium ion batteries with reliable and consistent performance. However, SK hesitated to expand its lithium ion battery business in full scale because of the absence of a captive market. SK had reserved the opportunity to become the 10th lithium ion battery company after LG Chemical and Samsung SDI. Instead, SK's chairman decided to move all the activities associated with batteries to the main company, SK Energy, to obtain the capability to invest in the battery business systematically. Prior to the decision on the full-fledged investment, SK Energy made a subsidiary called SK Mobile Energy (SKME), through which SK Energy carried out the pilot test on the lithium ion battery business per se to determine if, and by what method, battery business should be proceeded. Based on the results of the pilot test on the battery business through SKME, SK decided to focus on xEV batteries in December, 2007. Pursuant to the strategic decision on the battery business, SK Energy devoted its resources to the development of xEV batteries. After successful completion of the development stage, SK constructed the xEV battery plant in the western province of Korea and developed the working relationship with Continental of Germany. Now, the company identified as SK innovation, which was created to prepare for the full-scale competition in the world market, is leading the xEV battery business within SK group.

SK had to expand the battery business as expeditiously as possible, so the battery division of SK innovation hired many battery engineers in a short period of time, which may hurt SK innovation in the future. When we hire many people in a short period of time to establish the business foundation expeditiously, we cannot rule out the possibility of the inclusion of some bad apples. Battery business history shows that damage by the presence of rotten apples is much more serious than we think. For one thing, they drive away talented people from the organization. Bad apples know how to survive and their attitude is very contagious. Therefore, it is required that the battery business manager eradicate bad apples before any significant strategic action is to be taken. In the challenging business like lithium ion batteries, we need people who have the following three ingredients of success: desire to succeed, "Can Do" attitude and a team player. In retrospect, Korean battery industry was constructed brick by brick by a few battery people who have the ingredients of success mentioned above.

2) Late starters

The first company that had started lithium ion battery business in Korea was not LG or Samsung but Taeil Precision Inc. Polystor located in California, USA produced lithium ion batteries to supply lithium ion batteries to the US military market. Although there was no mass production lithium ion battery producers in the USA, there had been some battery companies that produced special purpose lithium ion batteries in small quantity. Polystor was one of them. Taeil Precision made 18650 cylindrical batteries using Polystor's electrodes and sent them back to Polystor. When Taeil Precision gained experience in lithium ion batteries through the battery assembly business, the parent company of Taeil Precision went out of business because of the financial difficulty. Taeil Precision was forced to shut down the plant, and all the battery engineers were moved to LG Chemical except one person who had signed with SK before LG's decision to accept Taeil Precision's battery engineers in 1997.

There existed another company that ventured to start the battery business prior to the market entry of Samsung and LG. It was Hanil Cement. Their main business is the production of the cement, which has no bearing on the battery business. In the 1990s Korea was the land of opportunity for Korean battery engineers in the USA, so many battery engineers came to Korea for the career opportunity. One of them who had gained some experience in Valence Technology of the USA approached Hanil Cement and convinced the president of Hanil Cement of the potentials associated with battery business. Since the president of Hanil Cement didn't quite understand the complexity pertinent to the lithium ion battery business, he welcomed the opportunity to enter into the battery business in the hope that the battery business could give Hanil Cement the opportunity to make a jump to the major league by competing with big companies like Samsung and LG.

Valence Technology, the LPB company of the USA, moved from San Jose to Ireland via Henderson, Nevada after its failure to provide the LPB samples to Motorola. Valence Technology changed the target battery from LPB to Bellcore's plastic lithium ion (PLI) batteries, and designed PLI packs for mobile phones that could compete with 18650 lithium ion cylindrical batteries. Valence Technology's strategy was to remove the dead space of lithium ion battery packs by using plate type PLI cells. At that time, two cylindrical cells were connected in series in the mobile phone power pack.

During the period of operation in Ireland, Valence Technology was contacted by Hanil Cement of Korea through a Korean battery engineer who used to work with Valence Technology to explore opportunities in Korean mobile phone market. After discussing the terms and conditions of the agreement, both parties signed the contract in 1996. Pursuant to the agreement, Hanil Cement formed a joint company called Hanil-Valence and started to construct the lithium ion battery assembly plant in 1996. Electrodes were to be supplied from Valence Technology. The Korean battery engineer who had introduced Valence Technology to

Hanil Cement led the battery business of Hanil-Valence. During the construction of a battery plant, the operating voltage of mobile phones was reduced to 3V from 5V due to the Motorola's development of a 3V power amplifier. The power amplifier of the mobile phone was the last component whose voltage should be reduced to 3V for the realization of 3V mobile phones. As a result, one cell power pack with a prismatic lithium ion battery became the standard pack in mobile phone market. PLI battery was too bulky to compete with one cell prismatic lithium ion battery power pack. After several years of struggle, Hanil-Valence was forced to shut down the PLI battery assembly plant. Hanil-Valence became the victim of the moving target, which is characteristic of the battery business.

10
:
Battery development

Although the 1980s was the era when the battery development was very active all over the world, Korean battery industry couldn't participate in this movement. Instead, it was lying dormant throughout the 1980s. The battery business was reserved exclusively for small firms in Korea, so most of the sizable commercial firms were not allowed to develop or produce batteries by law. As a result, batteries were outside the scope of the target business for most of the Korean firms. The 1980s was the dark age of the Korean battery industry.

After a long hiatus, Korea was beginning to understand the importance of the secondary batteries. As a matter of fact, it was Sony which made the Korean industry aware of the potentials associated with battery business. Overwhelmed by the Sony's success, LG and Samsung felt impelled to initiate secondary battery business. In addition, their participation in portable electronics business such as mobile phones and laptop computers gave Samsung and LG the opportunity to have an intimate knowledge on the necessity and potentials of the secondary battery business. Moreover, their strong desire to expand their marketing horizon enabled them to take part in the promising, but highly competitive market.

LG and Samsung had to start the battery development from the ground up. They had no choice but to resort to the outside source because there was practically no one within Korea who had the experience in the secondary battery business. In the early 1990s, Rocket Electric Co. produced alkaline manganese batteries and Sebang Battery Co. made lead-acid batteries. That's all there was in relation to the battery business in Korea. LG looked for help from the UK, and Samsung tried to search for the solution in the USA because both Samsung and LG thought western nations were the best source of knowledge in the battery development.

1) LG's development activities

(1) LG and AEA

In the early 1990s, LG of Korea signed the umbrella contract with AEA of the UK for the development of technology that could lead to commercial fruition in a perceivable future. In accordance with this contract, LG sent a high position delegate who had a good command of English to AEA for six months to select a development project. In response to this, AEA set up the branch in Seoul, Korea for the efficient pursuit of the development project. After the extensive discussions and scrupulous evaluation, both parties agreed to work together for the commercialization of LPB.

The basic idea of LPB is to make it possible to use high energy capacity lithium metal anode safely by adopting a polymer electrolyte. A polymer electrolyte is the polymer film that acts as both the electrolyte and a separator simultaneously, which makes the battery solid by the absence of the liquid component. But the fact that LPB can be operated only at elevated temperatures limited its application significantly. The 1990s was the decade of portable electronics such as mobile phones, laptop computers and the camcorder, which were called the 3C application. For the application of LPB in the 3C application, the operating temperature range was to be widened extensively. Both parties agreed to develop room temperature LPB and apply LPB to LG Electronics' laptop computers as an effort to extend the application area of LPB to the 3C market.

The senior managers of LG group wanted LG Electronics, an electronic company within LG group, to participate in the development project since LG Electronics was making laptop computers, but LG Electronics couldn't afford to allocate the resources to the development of LPB because it was tied up with the development of TFT-LCD (thin film transistor liquid crystal display), another hot topic at that time.

LG's next candidate was LG Chemical, the main company within LG group, but LG Chemical was resistant to take part in the LPB project as the research engineers of LG Chemical thought that it would be extremely difficult to make LPB working at low temperatures. In actuality, LG Chemical was interested in the battery business because it was in the process of changing the core business from polymer materials to the materials and devices associated with portable electronics. In this respect, the battery business was one of the ideal choices, but LPB did not coincide with LG Chemical's vision for the battery business. It was just too risky and chancy as the initial development project.

Next candidate was LG Metals, copper smelting and fabrication company. LG Metals accepted the challenge since LG Metals had been looking for a new business opportunity that could reduce its dependence on the copper smelting business. LG Metals wanted diversification in the business portfolio, so it agreed to undertake responsibility of room temperature LPB development in spite of obvious technical barriers.

Three year development project between LG Metals and AEA started in January, 1994. To begin with, training of LG members on the battery technology was conducted for one month at AEA in February, 1994. The way LPB development project was conducted was through technical meetings held every three months. Lithium polymer battery development project was essentially gradual technology transfer from AEA to LG. AEA, which is short for Atomic Energy Association, was a government funded institute, the headquarter of which was located in Harwell close to University of Oxford. The main research area was the development of atomic power including the design of a nuclear reactor. In fact, AEA had competed with Westinghouse of the USA for the acquisition of the contract for the construction of the nuclear reactor in Korea in the 1970s. In addition to the nuclear reactor, AEA was also strong in the development of the portable power like LPB. It is said that Harwell was the site of a headquarter and airplane base for the Allied Forces during World War II. After the war, British government turned this place into the atomic energy R&D center. Interestingly, AEA used an aviation shed as the site of laboratory and pilot plant.

LG Metals was supposed to pay a large sum of money to AEA every year for three years in return for the co-development of LPB. LG Metals wanted to have the right to stop the development project in midstream if the interim results didn't look promising. To reflect LG Metal's desire to have the right to withdraw from the agreement in midstream, the agreement was designed to determine whether or not LG Metals should continue the project at the end of each year.

Valence Technology of the USA was the leading company in LPB in the early 1990s. Valence Technology worked together with Motorola for the development of LPB for mobile phones. To reinforce the connection with Valence Technology, Motorola provided the development fund to Valence Technology, but unfortunately Valence Technology could not provide decent samples that meet Motorola's specifications. News of Valence Technology's failure to provide the working samples to Motorola made LG Metals aware of the fact that LPB might not be the solution for the future market. LG Metals wanted to switch from LPB to a more realistic alternative. At that time AEA had a good relationship with Sony thanks to the cathode patent. AEA suggested to LG Metals that LG Metals conduct lithium ion battery power pack business concurrently with LPB development, and arranged the meeting with Sony.

AEA hosted the meeting between LG Metals and Sony in Tokyo in December, 1994 and asked Sony to provide LG Metals with power pack technology for the pursuit of lithium ion battery power pack business, in which LG Metals was to supply the power pack to LG Electronics using Sony's 18650 cylindrical batteries, but Sony turned down AEA's offer in that since there were too many things Sony didn't understand about lithium ion batteries, it was too risky to provide lithium ion cells to other companies. One month after the unfruitful meeting between Sony and LG Metals, Sony surprised both LG Metals and AEA by proposing to develop lithium ion polymer batteries through a three-way development project. LG Metals and AEA agreed to conduct LiPB project.

The reason for Sony's proposal of LiPB development project is like this. Sony was the only battery firm that did not participate in the New Sunshine Project because of the Sony's policy that it should not collaborate with other companies. Sony must have felt left alone, which made Sony propose the development project of lithium ion polymer batteries to AEA and LG Metals to fill the void in the development strategy. The LiPB is an all solid lithium ion battery incorporated with a polymer electrolyte. Sony, LG Metals and AEA discussed development schedule and job allocations for a couple of months and reached the agreement, but to the surprise of all three parties, the senior managers of Sony rejected the proceeding of the project on the same reason as the case of New Sunshine Project.

AEA wanted to switch from LPB to lithium ion polymer batteries to maintain the viability of the development project even without involvement of Sony, and with this in mind, discussed the change of a project with LG Metals, but this discussion lasted for almost a year without any tangible results. The reason for the failure to reach an agreement in relation to AEA's intention to switch to lithium ion polymer batteries was that AEA couldn't come up with a convincing idea on how to develop lithium ion polymer batteries that had at least marketable performance.

In the meantime, LG's Chairman, Mr. Bon-Moo Koo, decided to move all the activities associated with batteries from LG Metals to LG Chemical as he wanted the main company to carry out the important project like lithium ion batteries for the aggressive full-scale investment. After receipt of the battery business from LG Metals, LG Chemical agreed to switch from LPB to lithium ion polymer batteries and continued the battery development project between LG Chemical and AEA until the end of the third year. Three year battery development project between LG and AEA came to an end in December, 1996. During the proceeding of a three year project with LG Metals and LG Chemical, AEA became the commercial company from the government-funded institute and changed its name to AEA Technology.

LG realized that development of the next generation battery without any battery business base would be meaningless, and changed the strategy from its focus on the next generation battery to the establishment of the battery business base. LG turned its attention from Europe to Japan for the sound acquisition of the battery business base.

(2) LG and Toshiba Battery

During proceeding of a LPB project with AEA Technology, LG Metals realized that such a next generation battery as a LPB might not lead to commercialization in a short period of time, so it decided to find a way to establish the battery business base in parallel with the development of LPB. LG Metals had some association with Mitsui Trading Company of Japan for the purchase of special chemicals, so LG Metals asked Mitsui Trading to arrange the meeting with Japanese battery companies for the collaboration in the battery business. In response to LG Metals' request, Mitsui Trading had introduced Toshiba Battery to LG Metals.

Toshiba Battery was the battery company that commercialized nickel metal hydride batteries in 1990. In fact, Toshiba Battery and Panasonic were the first companies to commercialize nickel metal hydride batteries in the world. Nonetheless, Toshiba Battery had been treated as a kind of battery business venture because unlike other Japanese battery companies, Toshiba Battery jumped to nickel metal hydride batteries without going through the conventional nickel cadmium batteries. Toshiba Battery wanted to compensate for this limitation, so it formed alliance with Varta of Germany and Duracell of the USA and called this alliance a 3C alliance. For the formation of the 3C alliance, Mitsui Trading had arranged the meeting with Varta and Duracell by capitalizing on the extensive international network. That's how Mitsui Trading had a close relationship with Toshiba Battery.

In a meeting with LG metals hosted by Mitsui Trading, Toshiba Battery agreed to supply the prismatic nickel metal hydride battery assembly line to LG Metals in 1995 and, more importantly, agreed to support LG Metals' lithium ion battery power pack business by providing lithium ion cells and the power pack assembly technology.

Before signing the agreement between LG Metals and Toshiba Battery, all the activities associated with batteries were moved to LG Chemical in April, 1996 pursuant to the LG Chairman's decision. After receipt of the battery business from LG Metals, LG Chemical signed the agreement with Toshiba Battery regarding a nickel metal hydride battery assembly and a lithium ion battery power pack business. LG Chemical constructed nickel metal hydride battery assembly plant using the equipment supplied from Toshiba Battery. Electrodes were supplied from Toshiba Battery on a regular basis and prismatic cells were to supply to Motorola for usage in the mobile phones.

At the time of the inception of the mass production of nickel metal hydride batteries, Motorola decided to discontinue the production of the mobile phones in which LG Chemical's prismatic nickel metal hydride batteries were to be used. All of a sudden, LG Chemical's target market had disappeared. Nevertheless, LG Chemical kept on operating the production line in the hope that it would be able to find a suitable market sometime soon. LG Chemical's hope turned out to be a forlorn hope. The storehouse ran out of space because of the inventory. As the inventory of batteries reached five million cells, LG Chemical shut down the plant and sold the equipment to Peacebay of China through Mitsui Trading. LG Chemical was not able to sell a single cell of prismatic nickel metal hydride batteries to any mobile phone companies.

The lesson LG Chemical had learned through prismatic nickel metal hydride battery business was the fact that the battery market was so changeable. As the battery market is a moving target, close attention to the change of the market is required not to make the same mistake LG Chemical has made in the nickel metal hydride prismatic battery business. LG Chemical realized how important it was to understand the market trend before any strategic decision was made. Based on the bitter experience related to the prismatic nickel metal hydride battery business, LG Chemical strengthened its marketing and product planning activities not to repeat the same mistake.

Toshiba Battery recommended LG Chemical to conduct the lithium ion battery power pack business prior to the lithium ion battery business for the understanding of the nature of the battery business. Pursuant to Toshiba Battery's recommendation, LG Chemical acquired lithium ion battery power pack technology from Toshiba and built a power pack assembly plant. It made power packs for mobile phones using cells supplied from Toshiba Battery and supplied to LG Electronics. This power pack business had continued until the lithium ion battery plant was completed in 1999. Thanks to the power pack business, LG Chemical was able to acquire the sales and marketing experience before the construction of the lithium ion battery plant. The battery business division of LG Chemical formed and grew the battery sales team when the battery business was in the development stage. When 18650 cylindrical batteries were made in the pilot plant, the battery sales team sold lithium ion cells in the after-market of Hong Kong, Taiwan and China.

At that time, LG Chemical's battery manufacturing skills were in the primitive stage, so electrolyte leakage, which is the most primitive quality problem, was the major problem. Some people in the after-market used to taunt the LG Chemical's cells by calling them crying cells. In the humid environment like in Hong Kong, Taiwan and the southern part of China, leaked electrolytes corroded a metal can quickly and left the red scar on the surface of the can. Seeing the corroded surface of the can, some people in the after-market said that LG Chemical's 18650 cylindrical batteries not only cried but also shed bloody tears profusely. Hence, dealers in the after-market used to say that LG Chemical's cells were shedding bloody tears once they were crying because the heart of LG Chemical was filled with sadness and sorrow.

Although LG Chemical's 18650 cylindrical lithium ion batteries were called crying cells, they were very popular in the after-market because LG Chemical's cells were the only lithium ion batteries available at the after-market because Japanese battery makers completely ignored the after-market. One day a truck filled with lithium ion cells made in the pilot plant of LG Chemical was stolen in an expressway rest area in Korea. Police failed to find any trace of a truck. The stolen batteries might have been consumed in the after-market. As we can see in the batteries robbery incident, lithium ion batteries were a high commodity in the 1990s because Japanese battery makers, the sole providers, supplied their lithium ion batteries only to the big portable electronics firms. It was not available to the public and the small firms.

2) Samsung's development activities

Samsung's battery business activity started at Samsung Electronics, the main company within Samsung group. Samsung Electronics acquired manufacturing technology of nickel metal hydride batteries from Ovonics of the USA and built the pilot plant in the early 1990s. The first target market was the cordless phones. Battery engineers of Samsung Electronics were confident that they could make nickel metal hydride batteries for the cordless phones in a short period of time. However, on the contrary to their expectation, they were not able to make a decent cell mainly because of the breakage of electrodes.

Ovonic's nickel metal hydride batteries were designed to stack the flat electrodes, but Samsung Electronics pushed to apply the winding process to make cylindrical batteries. Electrodes were cracked in the winding process because of the stiffness of the electrodes. In addition, battery engineers of Samsung Electronics had the misconception that battery manufacturing was basically an assembly process like electronic appliances. In fact, their confidence came from their observation that nickel metal hydride batteries had less than twenty parts. In the assembly business like electronic appliances, difficulty in the production technology increases in proportion to the number of parts. Acceptance of this logic implies that making a battery with less than twenty parts would be a cinch. However, no matter how hard they tried and no matter what stunt they pulled, they couldn't make jelly-rolls for the cells by the winding process. It was a real big blow to their ego.

Subsequent to all the futile attempts to make nickel metal hydride cylindrical batteries, Samsung group finally came to terms with the fact that the battery business needed more than just the assembly technology. Based on the bitter experience through nickel metal hydride battery business, Samsung group decided to move the battery business from Samsung Electronics to Samsung SDI. Samsung SDI produced the braun tube at that time. LG moved the battery business to a larger company for the enhancement of the investment capability. In contrast, Samsung moved the battery business from a large company to a right-size company for the better understanding of subtleness of the battery business.

Samsung's electronic business history is rather short as compared with LG. Samsung established the joint company called Samsung-NEC with NEC of Japan to acquire the braun tube technology in January, 1970. Samsung's Chairman, Mr. Kun-Hee Lee, had the firm belief that Samsung needed to learn about Japanese firms thoroughly to beat Japan as he thought that Japan had the best source of knowledge in the industry Korean companies focused on. As a result, Samsung is similar to Japanese firms in many respects. Morning gymnastics before starting work, greeting before the meeting, recording of minutes word for word and the self-examination meeting through reading by turns of the investigation report are good examples of Japanese influence on Samsung. In contrast, LG had absorbed the western way of working by working with the global consulting firms.

11
:
Lithium ion battery business

Since battery business had been reserved exclusively for small companies for many years in Korea, many Korean companies had to develop lithium ion batteries from scratch. They couldn't find any source of knowledge within Korea. In the situation like this, they were forced to find the source of knowledge and experience outside, and they had to find it as expeditiously as possible not to be left behind in the competition. Unlike other countries, lithium ion battery had to act as the seed battery business in Korea because they had practically no foundation of the secondary battery industry. In essence, new industry of a secondary battery was beginning to emerge in the 1990s in Korea, about thirty years after Sanyo's introduction of sealed nickel cadmium batteries to the portable electronics market.

1) Beginning of the battery business

(1) Technology inflow

In 1996, surprisingly enough, Sony released some battery engineers who had played an important role in commercialization of lithium ion batteries as a result of conflict between battery engineers and the senior managers. On the contrary to the battery engineers' desire to expand the battery business aggressively, senior managers wanted to expand the lithium ion battery business with extreme caution and thoughtful consideration at Sony for obvious competitive reasons. Sony's intention was not to become a battery manufacturer like Sanyo and Panasonic. Sony's emphasis had been traditionally on novel portable electronics.

Korean battery companies like Samsung SDI and LG Chemical heavily benefited from the technology inflow from Japan. Japanese battery engineers became important assets to Samsung SDI and LG Chemical. In addition, top-notch engineers in Korea gathered around lithium ion battery business because they thought they would be able to find a suitable career opportunity in the battery business. Both Samsung SDI and LG Chemical constructed lithium ion battery pilot plants and developed mass production technology competitively. Financial crisis that had never experienced before hit Korea in 1998. It hit so hard that it changed the fundamental social structure of Korea completely. Different face of Korea emerged after the closure of financial crisis. During the financial crisis, all the investments were cancelled since most commercial firms including the government were cash-poor. Nevertheless, LG Chemical and Samsung SDI continued their investment in the lithium ion battery business regardless.

LG Chemical needed to strengthen its capabilities in the electrode production and equipment design to compensate for the intrinsic weakness as a chemical company. LG Electronics, a sister company within LG group, had a strong technological background in video tape manufacturing, equipment design and automation. LG's Chairman sent the equipment designers of LG Electronics to LG Chemical to work together with battery engineers side by side from the beginning stage of the battery development. In so doing, LG Chemical was able to design the battery plant on its own after the first battery plant. In the first battery plant, LG Chemical installed the Japanese equipment to reduce the lead time. As a result of aggressive investment and timely collaborative work, LG Chemical was able to beat Samsung SDI in the competition on which company would be the first company to commercialize lithium ion batteries in Korea. LG Chemical held ceremony for completion of lithium ion battery plant in October, 1999. LG Chemical became the 8th lithium ion battery company after Hitachi-Maxell of Japan. President of Toshiba Battery attended the opening ceremony to make a speech commemorating the completion of the battery plant.

At the present time nobody would deny that Samsung SDI is a battery company, but Samsung SDI's main business was the display like braun tube in the 1990s. Battery business was just the addition to the display business, which delayed the investment of lithium ion batteries in competition with LG Chemical. However, availability of the lithium ion battery technology in Japan urged Samsung SDI to move fast not to get left behind in the competition of the secondary battery business. Samsung SDI's strategy was to capitalize on its experience in the display business as an electronics company. It installed the fastest coating machine and the largest dry room for the battery assembly line. Samsung SDI collaborated with Umicore of Belgium to develop LCO cathode materials, and applied aluminum coating on the powder surface to obtain the extended cycle life. It also designed a new CID for the cylindrical batteries and a side vent for the prismatic batteries for the price competitiveness.

After completion of the development stage, Samsung SDI constructed the mass production plant and held the opening ceremony of the battery plant in July, 2000. Samsung SDI became the 9th lithium ion battery company after LG Chemical. Samsung and LG had been in rivalry in many business areas for a long time. Lithium ion battery was just one of them. Six sigma movement came from Motorola of the USA to Korea through Sony in 1990s. Both Samsung SDI and LG Chemical were avid believers of six sigma methodology. Six sigma movement became the foundation of their battery business.

For the successful pursuit of a new business, we need three factors: manufacturing technology, materials and the equipment. Samsung SDI and LG Chemical had purchased manufacturing equipment and battery materials from Japan in their first battery plant to shorten the commercialization time frame. In actuality, Japan had been securing its lithium ion battery technology by restricting access of materials and equipment firms to foreign battery companies like Samsung SDI and LG Chemical. The restriction was lifted in 1996. Although Korea was successful in producing lithium ion batteries in a large scale, its dependence on Japanese materials and equipments was dangerously high. Korea had to reduce its dependence on imported materials and equipment for the competition with leading Japanese battery makers in the world market.

On account of a long history of Korean government's appointing battery business exclusively for small firms, infrastructure of the battery-related business was very weak. As a first step to establishment of infrastructure, Samsung SDI and LG Chemical formed Korean Battery R&D Association in 1997. As the battery industry expanded, Korean Battery R&D Association grew to be Korean Battery Industry Association. Korean Battery R&D Association provided the venue where battery makers meet together with materials and equipment makers to exchange ideas and information with each other, which facilitated the development of battery materials and the equipment. These systematic efforts finally paid off. Battery maker's dependence on imported materials and equipment was beginning to be reduced.

(2) Battery plant operation

In a new business like a lithium ion battery, operator's knowledge and experience determine how smoothly the battery makers can mass produce the battery. More often than not, this critical factor in relation to the mass production is ignored. LG Chemical's decision to mass produce the battery is based on the acquisition of a purchasing order from LG Electronics and a laptop computer pack maker of Taiwan.

LG Chemical started to make lithium ion cells in the plant in October, 1999. After 28 days aging, cells produced in the plant was tested to determine if the cell performance was same as that of the cells made in the pilot plant. Nobody within LG Chemical worried over something coming up. However, on the contrary to the expectation, the cycle life dropped prematurely. LG Chemical was not able to supply lithium ion cells because cells they produced failed to meet the cycle life requirement. The mass production line was forced to be stopped in December, 1999. LG Chemical's nightmare to have failed to sell a single cell of the nickel metal hydride battery back in 1997 revived. Devastated, research engineers who designed the battery and the mass production line were sent to the battery plant to investigate the cause of the cycle life problems. They used all sorts of methods including six sigma tools to figure out what caused the premature drop of cycle life. In spite of their efforts, six months had passed without finding any clue. One day, they acknowledged some abnormality in the statistical data of the roll pressing process. They found that the operator of the roll pressing process didn't operate the rolling machine as indicated in the recipe. Excessive roll pressing was found to be the cause of premature drop of cycle life.

LG Chemical constructed the second battery plant for the production of the proprietary pouch batteries. Unlike the first battery plant, it designed most of the equipment on its own with equipment designers from LG Electronics. LG Chemical was really keen not to repeat the same mistake it had made in the first battery plant. The battery business division even introduced design for six sigma (DFSS) in the design stage. Battery engineers of LG Chemical also trained the operators intensively. However, in spite of all the preparations and efforts, they could not sell the pouch batteries for six months after the inception of the operation because of the swelling of cells. In the turmoil of this, LG Chemical found out the cause of swelling of cells by accident. The operator in the electrolyte injection process cleaned the electrolyte injection nozzle with alcohol. Hydroxide ion (OH^-) in alcohol reacted with the electrolyte to form hydrogen gas, which made the cells swelling severely. LG Chemical's mass production of the battery was stabilized two years after the inception of the mass production mainly due to the improvement of operator's skills and knowledge. In contrast to LG Chemical, Samsung SDI's mass production proceeded according to the plan. Samsung SDI had many engineers from Japan who had a great deal of hands-on experience in the battery production and quality control. That made a big difference.

2) xEV batteries

(1) The future battery project

The gap between Korean and Japanese battery companies was almost ten years in terms of the start-up year of the lithium ion battery business. As far as the battery development is concerned, the gap is enormous. For the followers to catch up the leader, they should do more than following in a leader's footsteps. Either they develop leap-frog technology for the quantum jump or they should move ahead of others in an emerging market. Otherwise, the gap with the leading group will become bigger and bigger.

In the 1980s and 1990s, Korean firms diversified business recklessly and competitively without regard to their technological background, which resulted in financial crisis in 1998. As a remedy to this deep-seated intrinsic problems, Korean government enforced merging between firms with the same business, which was called the Big deal. LG group, an electronics and chemical conglomerate, and Hyundai group, an automobile conglomerate, competed with each other for the semiconductor business in the Big deal. Most of people in Korea thought that semiconductor business would be merged to LG because LG had a strong groundwork for the electronic business and a long history of electronic appliances. In fact, it was LG, which was originally called Goldstar, that made the first transistor radio in Korea. However, contrary to the expectation, semiconductor business was merged to Hyundai group. LG was dumbfounded by the Korean government's completely unexpected decision. However, that's what had happened and LG had no choice but to accept the ramifications of the Big deal.

As LG had lost the core business associated with electronic business, senior managers of LG felt compelled to establish the strategy to revitalize the remaining businesses. With this in mind, senior managers of LG decided to pursue a project in which they find a way to improve the competitiveness of the remaining core businesses. With the help of BoozAllen & Hamilton of the USA, LG group conducted the project that determined the core competence (CC) strategy in 1999. The director who was in charge of the Big deal competition led the CC project because he was the one who could see the global picture of the main businesses within LG group although he failed to win out over Hyundai in the Big deal competition.

Lithium ion battery business of LG Chemical became the first project of LG group's CC project. Since CC project had been conducted before the completion of the first lithium ion battery plant, CC project seemed to have leeway to determine if, and how, the lithium ion battery business should be proceeded. BoozAllen & Hamilton had no knowledge and experience on the lithium ion battery, so it subcontracted to a one man Japanese battery consulting firm for the expert advice. After investigating the battery market and the competition dynamics, the CC project team recommended LG Chemical to find a new battery business opportunity in an emerging market in parallel with closer attention to the bottom line of a small lithium ion battery business because the CC team thought that lithium ion battery business had already entered the stage of cost competition in which Japanese battery makers excelled. Some senior managers within LG group even argued that construction of the lithium ion battery plant should be stopped. However, LG decided to go ahead with the construction of the lithium ion battery plant according to the original schedule despite the adverse business environment. As far as a newcomer like LG was concerned, lithium ion battery business was the seed business of Korean battery industry. Some seeds had to be spread regardless, and the seed called the lithium ion battery was a promising one for the future battery business, considering. Samsung thought the same way. As a follow-up to the core competence project, LG Chemical identified emerging markets it should strive for. The first project was the xEV batteries and the second one was the electrodes of fuel cells in the order of priority. In this way, xEV batteries became the number one future project for LG Chemical in 2000.

Samsung had to give up the automobile business as a result of Big deal. Unlike the semiconductor business of LG which had acted as the main business within LG group, Samsung's automobile business was a new business, so the damage due to the loss of automobile business was not so devastating as that of LG. Nevertheless, senior managers of

Samsung were beginning to feel that they needed to take the systematic approach to the future businesses. Pursuant to this plan, Samsung SDI carried out the project of determining the future business with the help of Nomura Research Institute (NRI) of Japan in 2002. Based on the investigation on the new business associated with Samsung's long-term vision, Samsung SDI chose mobile fuel cells for laptop computers and xEV batteries as the future business project in the order of priority. Therefore, mobile fuel cells for laptop computers became the number one future project for Samsung SDI.

The result of the CC project coincided with Samsung SDI's desire to be a total energy provider. The choice of mobile fuel cells for laptop computers instead of xEV batteries as the future business reflects Samsung SDI's intention to be a total energy provider by combining small lithium ion batteries with solar cells and fuel cells. Samsung SDI wanted to jump to a total energy provider without going through xEV batteries to shorten the time frame. In spite of the favorable bottom-line of the display business in the early 2000s, Samsung SDI was desperate to transfer from the display to the energy firm swiftly because the braun tube was disappearing in the market much faster than Samsuing SDI expected and PDP business was running a huge deficit in competition with TFT-LCD. Hence, senior managers of Samsung SDI needed to come up with something big to replace the braun tube business as quickly as possible. That's why they chose to pursue mobile fuel cells for laptop computers in spite of the obvious technical barriers.

After determination of the number one future project, Samsung SDI put most of the available resources to the development of mobile fuel cells instead of xEV batteries. They hired over 120 people in a short period of time and constructed a pilot plant. However, their efforts didn't pay off. Problems in relation to the mobile fuel cells were too big for Samsung

SDI to tackle with. Reaching a dead end, Samsung SDI decided to put a stop to the development of mobile fuel cells for laptop computers in 2006 after four years aggressive R&D investment. They failed to find a magic solution like so many other previous aborted fuel cells projects. Samsung SDI killed two birds with one stone, but unfortunately the birds they killed belonged to Samsung SDI.

In conclusion, Samsung SDI had lost the chance to grow its xEV battery business because of the strategic blunder. The most important part of the business management is to get the priority straight. Any mistakes in the priority decision would hurt the business badly and, to make matters worse, would leave the permanent scar.

(2) LG's activities

High Energy Technology (HET) of the USA was once known as the company which had a very promising manufacturing technology among Bellcore's plastic lithium ion (PLI) battery licensees. HET provided PLI cells to LG Electronics of Korea for the mobile phone application together with its collaboration with Hyundai Motor of Korea for the development of xEV batteries through Compact Power, one man company in the USA. In the midst of all these activities, the parent company decided to give up HET's battery business on account of the financial difficulty. HET and Compact Power wanted to maintain their viability within the battery industry by passing their knowledge and experience associated with xEV battery business on to other companies. With this in mind, LG Chemical was approached by HET Korea and Compact Power to set up a joint company identified as Compact Power Inc. (CPI) for the marketing and power pack development. After several month's comprehensive talks, LG Chemical decided to establish CPI in Colorado. The opening ceremony of CPI was held in October, 2000. On behalf of

LG Chemical, the battery business manager participated in the ceremony to exhibit his strong support to CPI.

LG Chemical's involvement with plastic automobile parts business made it easy to decide on the establishment of a marketing branch close to the main market. Hyundai Motor also welcomed LG Chemical's decision with the gesture to make an equity investment in CPI, which was aborted by the management change of Hyundai Motor. CPI succeeded in penetrating into Hyundai Motor's xEV market using LG Chemical's pouch batteries in 2002. Through this development project, LG Chemical formed collaborative ties with Hyundai Motor. Afterwards, CPI was moved from Colorado to Detroit to devote its efforts exclusively to marketing rather than pack development. CPI was established by the LG Chemical's investment of three million dollars capital. Some portion of the equity was transferred to the member of Compact Power who used to work with HET as a token of LG Chemical's acknowledgement of their experience and knowledge on xEV batteries, as a result of which CPI became a joint company of LG Chemical and Compact Power.

Unlike electronic companies, chemical company like LG Chemical is extremely cautious in the investment. Moreover, creation of a joint company is a subject that most companies treat with extreme caution and thoughtful consideration for the obvious competitive reasons. As for LG Chemical, it was a big decision to make a joint R&D firm away from Korea, however small. For this reason, the CEO of LG Chemical was reluctant to approve of setting up a joint company in the USA because he thought that there were so many things he didn't understand and, more importantly, he didn't want to be blamed for something he had no control over in the future. In addition, the first lithium ion battery plant he had approved of was not operable because of the quality problem. The CEO of LG Chemical wanted to clear up all the ambiguities associated with CPI before the approval of a joint company.

Approval request from the battery business division was rejected by the CEO with a memo of questions regarding the xEV battery business. On receipt of the rejected approval request form, battery business division submitted the revised approval request with the answers to the CEO's questions. To-and-fro movement of approval requests between the CEO and the battery business division was repeated five times until the CEO of LG Chemical ran out of the questions. Formation of a joint company was finally approved by LG Chemical. The next step was to get the approval from the chairman of LG group. This approval step was easier than the process of LG Chemical because all the ambiguities relative to the xEV battery business were cleared up during the approval procedure of LG Chemical. After the acquisition of chairman's approval, CPI finally became a joint company of LG Chemical and Compact Power in October, 2000.

According to the business plan of CPI, no additional investment was required to operate CPI because the operating expenditure would be small and it was expected that CPI would receive enough research fund from the outside source. However, one year after the establishment of a joint company, CPI ended up with using up all the money invested by LG Chemical. CPI put more focus on the development of power packs rather than marketing, so it hired research engineers and installed the power pack development equipment. The use of its capital in the investment of fixed asset instead of the operating expenditure made CPI cash-poor. LG Chemical stepped in to get things straight. Vigorous arguments continued between CPI and LG Chemical as to how to operate CPI for several years. Finally, there was a management shake-up at CPI, as a result of which CPI was owned solely by LG Chemical. LG Chemical moved CPI from Colorado to Detroit, the center of the automobile industry, and focused on marketing. In this way, CPI was reborn as a marketing branch of LG Chemical.

All the people related to the establishment of CPI and xEV battery development project left LG Chemical in December, 2001. History associated with LG Chemical's xEV battery project was buried into oblivion. Ignorant of the history related to the establishment of CPI, a newly appointed president and a battery business manager found it hard to justify the operation of CPI. They thought that it would be beneficial for LG Chemical to focus on a small lithium ion battery business used in portable electronics. They were under the impression that LG Chemical couldn't afford to prepare for the future market like xEV batteries, which aggravated the relationship between LG Chemical and CPI. During this period, LG Chemical was losing the competition with Samsung SDI in spite of the head start. Capitalizing on the conflict of the xEV batteries in LG Chemical, Samsung SDI hired an engineer from Honda of Japan and signed the development contract with Ford to compensate for the belated start in 2005. Samsung SDI had recovered from the field incidents of laptop computers, but LG Chemical was right in the middle of the field incidents of Apple's laptop computers in 2005.

(3) Samsung's activities

Pursuant to the consulting result from Nomura Research Institute, Samsung SDI formed a development team of xEV batteries in the central R&D center located close to Seoul, capital city of Korea. The development team of small lithium ion batteries, which is the main force of battery development, was positioned in the battery plant located in the middle part of Korea. In general, many research engineers in Korea wish to work in the metropolitan Seoul because close proximity to the political, economical and cultural center of Korea enriches their lives. That's why

Samsung SDI rarely transfers research engineers from the battery plant to the central R&D center because once there is a movement from the plant to the central R&D center, there can be uncontrollable demand to move from the battery plant to the metropolitan area like a domino effect. If that happens, it's only a matter of time that the battery development team at the plant wobbles and crumbles.

After formation of the xEV battery development team, senior managers of Samsung SDI were keen to search for someone who had the capability to lead the xEV battery team from Hyundai Motor, but they couldn't find a person who fitted the bill. So, as an alternative measure, Samsung SDI had director in the battery development team who got recruited from LG Chemical a few years back head up the xEV battery development team because he had the experience of establishing an infrastructure of xEV battery business at LG Chemical.

When Samsung SDI started the development of xEV batteries in 2004, the environment surrounding the xEV batteries was not so favorable in Samsung. For one thing, loss of automobile business in the turmoil of Big deal made many high position senior managers within Samsung group allergic to the word an automobile or a vehicle, so development of xEV batteries might be in danger of antagonizing many decision-makers in Samsung group. Most of directors in the headquarter of Samsung SDI felt the same way. Only the president of Samsung SDI thought differently. He thought that if LG moves toward xEV battery business, Samsung should move as well.

The xEV battery team members needed to be blended in to maintain the viability within the system of Samsung SDI, so they coined acronym the HPL battery, which is short for the high power lithium ion battery, to replace the term of the xEV battery. True identity of the xEV battery had been concealed by the ambiguous term until the HPL battery team leader announced that he and his team members would work together with Ford for the development and the eventual commercialization of xEV batteries. Because of the invisible wall between the xEV battery and the small lithium ion battery development team in terms of the exchange activities of battery engineers, the experience and knowledge of the battery development team could not be utilized. The xEV battery team had to hire battery engineers mainly from Japan.

After signing the xEV battery development contract with Ford, the HPL battery team managed to find its true identity as the xEV battery team and built the pilot plant for the systematic development of xEV batteries. However, staff members of Samsung SDI's headquarter wanted to concentrate on mobile fuel cells instead of distributing the resources to both mobile fuel cells and xEV batteries. Argument between the xEV battery team and staff members as to the justification of the development of xEV batteries continued even after the president's official decision to develop xEV batteries for commercialization. In general, the argument stops and the consensus is reached when one of them leaves the organization. Director of the xEV battery team was forced to leave Samsung SDI in spite of the protection of the president who recruited him from LG Chemical, and the head of R&D center was moved to Samsung Fine Chemical in January, 2007. After moving to Samsung Fine Chemical, he led the R&D center and laid groundwork for the development of cathode materials for xEV batteries. Afterwards, Samsung Fine Chemical collaborated with Toda of Japan and formed a joint company called STM in May, 2011. STM makes nickel base cathode materials for the Samung SDI's battery division. Now STM is a member of Samsung SDI.

3) xEV battery market

(1) General Motors

General Motors (GM) had focused on the development of fuel cell vehicles since senior managers of GM thought that xEVs wouldn't last long and the fuel cell vehicle would be the ultimate solution. But as the project progressed, they were beginning to realize that they were wrong about the strategy on the future vehicles. Finally, GM decided to discontinue the research on the fuel cell vehicles and switched to xEVs, but the gap with HEV leaders was too large. As a way to turn the situation around, GM chose plug-in hybrid electric vehicles (PHEVs) as the target vehicle. Prior to GM's interest in the PHEVs, people in the automobile industry already knew the concept of the PHEVs very well, but the need for charging was a key problem confronting PHEVs for the lack of charging facilities and the inconvenience pertinent to charging.

The PHEV is the combination of an EV and a power assist HEV. When we switch the power on, the PHEV is driven by an electric motor, which is called an EV mode. After the energy capacity of the batteries drops to a certain level, a driving mode is changed from an EV to a power assist HEV mode. Another feature of a PHEV is the charging of batteries from outside. Because of the inclusion of an EV mode, energy capacity of the batteries jumps up to over 20Ah from 5Ah. GM decided to cooperate with LG Chemical in the development and commercialization of PHEVs in 2007. GM and LG Chemical, both of which had a very difficult time in 2006, endeavored to use PHEVs as a means to recover from the slump in their business. In this respect, they met with each other at the right time.

Toyota changed its initial plan to construct the lithium ion battery plant and built additional nickel metal hydride battery plant instead in 2006. Toyota's unexpected change of a plan made many people pertinent to xEVs have a second thought on the xEV business. They felt that the xEV business was on the decline. In the atmosphere of negativity surrounding the battery market, the joint project of GM and LG Chemical on the PHEV battery changed the flow in 2007. It galvanized the HEV business and, more importantly, it made European automobile makers that were so resistant to take part in the xEV business take actions. Therefore, the joint project between GM and LG Chemical will be arguably recorded as one of the most important events in the xEV history.

(2) Ford

Mr. Ford, Chairman of Ford, is a well-known environmentalist, which explains why Ford exhibited so much interest in xEVs. Thanks to the continuous support of the leader, Ford belonged to top three makers in HEVs. Sanyo of Japan had supplied D size (diameter 34.2mm, height 61.5mm) nickel metal hydride batteries to Ford. In the beginning, Sanyo provided power packs to Ford for the installation in Ford Escape. This battery was called Gen 1. After acquisition of power pack technology, Ford made its own power pack using the D size nickel metal hydride cells from Sanyo. This battery was called Gen 2. Gen 3 was to be lithium ion batteries. Ford had been searching for lithium ion battery makers other than Japanese companies with the aim of reducing its dependence on Japanese batteries and power pack components.

At that time, Samsung SDI was developing cylindrical and prismatic lithium ion batteries for xEVs concurrently with searching for a strategic partner. Members of xEV battery development team of Samsung SDI knew that unlike small lithium ion batteries for mobile phones and laptop computers, market-oriented research was necessitated in xEV batteries. With this in mind, they were beginning to meet with automobile makers in Europe and the USA to introduce their strategy and development plan.

In the midst of the tour to automobile companies in the world, Samsung SDI's xEV battery development team found a potential partner, Ford, which exhibited the interest in the Korean battery companies for the mutual benefit. Both parties had discussed the development project that would lead to commercial fruition throughout 2005. In the meantime, Samsung SDI devoted its resources to the mobile fuel cells for laptop computers, Samsung SDI's number one future project, and because of this, the administration group of Samsung SDI didn't like the idea of pursuing collaboration project with a major automobile company like Ford with the aim of commercialization.

The xEV battery development team and the administration group locked horns with each other over the xEV development project with Ford. To resolve this conflict, president of Samsung SDI held an official meeting in November, 2005. After hearing their debate, president of Samsung SDI decided on proceeding the xEV battery project with Ford on the grounds that Samsung SDI had no alternative. The approval from Ford's senior managers was also needed along with the already acquired Samsung SDI's approval. Samsung SDI's development team obtained the approval from Ford's senior managers after presentation of the strategy and development plan to senior managers of Ford thanks to the strong support from the battery team of Ford in March, 2006. Finally, the xEV battery development project between Ford and Samsung SDI began, with Ford picking up most of the tab.

The purpose of the joint development project was to develop lithium ion batteries which eliminated all the technical barriers. After development of commercially viable batteries in terms of battery performance and cost, Ford and Samsung SDI were scheduled to find the market together for the purpose of expanding the xEV market. The goal of the joint project was to construct the xEV battery plant using the technology developed through development project. This project enabled Samsung SDI to catch up LG Chemical which had a head start by more than three years. For a short period of time, Samsung SDI took the lead in a neck and neck competition with LG Chemical despite three years belated start. However, this lead didn't last long.

In the second half of 2006, Samsung SDI formed a marketing department and evaluated the feasibility of the xEV battery project. Marketing department submitted gloomy outlook to the senior managers. Pursuant to the marketing department's forecast, president of Samsung SDI decided to downscope the xEV battery project along with the restructuring of the xEV battery development team despite Ford's strong protest. Samsung SDI let the director who was in charge of the Ford project leave Samsung SDI, and reduced the size of the development team to a minimum and had the director of mobile fuel cells development team control the Ford project in January, 2007. The relations between Ford and Samsung SDI had been turned for the worse in 2007.

The xEV market was beginning to be reactivated after GM's announcement to collaborate with LG Chemical in the area of PHEVs. Business outlook for the xEV battery changed for the better. Samsung SDI's marketing team shifted its ground on the xEV battery business and provided the forecast reflecting the optimistic atmosphere surrounding the xEV business to the president in 2007. Based on the optimistic marketing report, Samsung SDI finally decided to focus on the xEV batteries. The xEV battery was beginning to emerge as the important project within Samsung group. In fact, the xEV batteries became the important future project in Samsung group as well as in Samsung SDI when president of Samsung SDI was promoted to a vice chairman and moved to the Chairman's office, where he was in charge of planning the future business of Samsung group.

Since Samsung SDI had the braun tube plants in Europe, it not only had good networks in Europe but was more familiar with European companies than the US companies. After a lengthy conflict with Ford, Samsung SDI formed an alliance with Bosch of Germany, forming a joint company called SB LiMotive (SBL). After completion of the development stage, SBL constructed the xEV battery plant in the southern province of Korea. However, the xEV business didn't progress as smoothly as SBL and Samsung SDI anticipated. Their joint business movement stopped abruptly on account of the accumulated loss. In the operation of SBL, the marketing team leader of Samsung SDI became the first CEO, followed by battery sales director of the small lithium ion battery business. Both of them didn't have the technical background. The appointment of persons without any technical background as the CEO of a joint company was based on Samsung SDI's misconception that the design of batteries was completed, but the battery design was far from complete. SBL should have been a joint R&D company which focuses on the completion of the cell design rather than sales of the batteries.

Most of automotive parts companies produce lead acid batteries for automobiles. In other words, the battery is one of their core business. Once the company like Bosch acquires the xEV battery technology, it can grow the battery business in a short period of time by taking advantage of its strong network and marketing channels. If automotive parts companies compete with traditional battery companies in the xEV market, chances are traditional battery companies will be likely to be outperformed and outsmarted. Therefore, when the battery companies are to collaborate with automotive parts companies, special caution should be exercised not to be overexposed.

(3) Hyundai Motor

Hyndai Motor used to develop the fuel cell vehicle as an ultimate future vehicle like GM. In fact, Hyundai Motor did not have to worry about California mandate as Hyundai Motor didn't belong to top seven makers in the 1990s. However, Hyundai Motor's merging with Kia Motor of Korea as a way to rectify the financial crisis of Korea in the late 1990s increased the likelihood of Hyundai-Kia's belonging to the top seven makers, so Hyundai Motor was beginning to take the xEV business seriously.

Hyundai Motor has been maintaining its working relationship with LG Chemical since 2002. SK innovation also became a partner of Hyundai Motor through the development project. However, Hyundai Motor has been reluctant to use Samsung SDI's batteries in spite of Samsung SDI's persistent effort to develop the working relationship with Hyundai Motor mainly because Hyundai Motor considers Samsung as the potential competitor because of the Samsung's aborted entry into automobile business back in 1990s.

Hyundai Mobis, which belongs to Hyundai Motor group, is the automotive parts firm like Bosch, Johnson Control and Continental. Like other automotive parts firms, Hyundai Mobis wished to have a much larger role in the xEV battery industry. With this in mind, Hyundai Mobis developed LFP batteries with Hanwha Chemical in 2011-2012. Hyundai Mobis wanted to develop LFP batteries which could compete with NCM batteries in terms of the energy density by taking advantage of the high safety features of LFP batteries. However, the technical barrier was too high to surmount.

Hyundai Mobis wanted to reduce its dependence on Hyundai Motor market, so senior managers of Hyundai Mobis were keen to find a way to move independently of Hyundai Motor. That's why they developed the xEV battery to acquire the core technology of xEVs. Hyundai Mobis' plan was either to sell its own xEV battery power packs to numerous car makers or to make its own electric vehicles without collaboration with Hyundai Motor. However, Hyundai Mobis' plan was thwarted by choosing a wrong partner. Hyundai Mobis got victimized by Hanwha Chemical's abrupt strategic decision to stop all the activities associated with lithium ion batteries in December, 2013.

12
:
Battery materials

The battery business doesn't belong to the realm of the material business. Nevertheless, by the introduction of the lithium ion battery which was referred to as the materials-dominant battery in the early 1990s, materials technology played an important role in the battery business. Competition dynamics of such battery makers as Sony, Sanyo and Panasonic back in the 1990s would give further insight into the nature of lithium ion battery business. In their competition for the market, new materials enabled one of them to take the lead in the cell performance and cost at the critical points. Sanyo, which had a strong background in materials technology, had led the lithium ion battery market by providing materials solution whenever the battery industry encountered technical difficulties. Sanyo developed the artificial and natural graphite and the introduction of compatible electrolytes and new manufacturing process accruing from the use of new anode materials.

Although materials technology acted as the solution provider at critical points for leading companies like Sony, Sanyo and Panasonic, battery production technology determined the competition in the portable electronics markets mainly because of the relatively short life cycle of cells. In small lithium ion batteries of mobile phones and laptop computers, the version-up of batteries, which is the energy capacity increase of the cells, occurred every ten months. In this fast moving market, improvement in cell design and manufacturing technology is the most effective way of winning the competition. In this regard, production technology became an essential part of the competition, in which Japanese battery makers excelled through their long history of battery business.

We are seeing the market for batteries change from potable electronics to xEVs rather slowly but with obvious assurance. Difference between portable electronics and xEV battery market can be found in the life cycle of batteries and the long approval time. Longevity of the battery model enables us to shift the focus from manufacturing to materials technology since battery makers not only have sufficient time to adopt new materials in their batteries but find it easy to enhance the cell performance through the adoption of new materials rather than the cell design. Therefore, battery materials are beginning to emerge as the critical factor to the competition of xEV battery business. Unlike materials used in other fields, it is essential that those who have strong background in the battery technology should develop battery materials to come up with workable materials. Otherwise, materials and battery technology wouldn't mingle with each other to make a homogeneous technology.

There is no hard and fast rule to classify the battery materials, but it is more convenient to classify the battery materials based on the battery manufacturing process. Battery manufacturing process is made up of the electrode making and the assembly process. Battery materials used in the electrode making and assembly process are called electrode and assembly materials respectively. Hence, electrode materials include active materials of cathode and anode, the current collector, binder and carbon black whereas the assembly materials encompass a separator and the electrolyte. Battery parts were excluded in this classification.

1) Electrode materials

(1) Cathode materials

Active materials of cathode and anode, the electrolyte and a separator are four major components in lithium ion batteries. Battery makers are active in the development of cathode materials even though they are not involved in the production of materials since cathode materials are instrumental in determining the cell performance. LCO became the major cathode material in the 1990s since Sony's battery engineers applied this material successfully to their lithium ion batteries. Interestingly enough, Sony's choice of LCO over other cathode materials was not based on the investigation on the pros and cons of each material. Rather, Sony's choice stemmed from its experience and knowledge of cobalt metal acquired in the time when Sony was developing samarium cobalt (Sm-Co) permanent magnet.

In the late 1980s there were three development groups in the cathode materials of lithium ion batteries: Sony of Japan, Moli Energy of Canada and Bellcore of the USA. Moli Energy developed lithium nickel oxide because the battery engineers of Moli Energy were familiar with nickel. Canada is known as the center of nickel production. On the other hand, the USA is strong in the primary battery like alkaline manganese batteries. Battery companies like Eveready and Duracell have been leading the alkaline battery market for a long time. Cathode materials of alkaline manganese batteries are manganese dioxide. Bellcore opted for spinel manganese oxide (LMO) as the target cathode material because battery engineers of Bellcore had a good understanding of manganese oxide. They all seemed to have adopted familiar materials as their cathode materials. Their selection is not based on the investigation of strength and weakness of each material. Familiarity to the materials turned out to be a primary criterion to determine the cathode materials of lithium ion batteries.

LCO is easy to manufacture, for it is made by simple solid state reaction and has a simple primary structure in powder morphology. These features have the advantage of giving reliable and consistent electrical properties regardless of the production method. In other words, unlike other cathode materials, the electrical properties and other pertinent characteristics of LCO are not sensitive to the manufacturing method. Although Sony didn't choose cathode material based on the scrupulous investigation on the property and manufacturing method, Sony's choice on the cathode material turned out to be an excellent one.

LCO has the layer structure, in which cobalt layer is sandwiched by two oxygen layers to form the cobalt oxide structure unit and lithium layer is sandwiched between these structure units. Owing to the layer structure, powder surface of LCO has the polar nature: positive for cobalt and lithium metals and negative for oxygen. When the electrolyte is contacted by the cathode powder, polar electrolyte molecules are attracted to their opposite charge sites on the powder surface, the electrolyte molecules being disintegrated. Disintegration of the electrolyte results in the decline of the cycle life. Diffusion barrier is to be formed by coating the surface of the cathode materials with metals to prevent the direct contact of the electrolyte with the cathode powder surface. In the beginning, aluminum was used as the coating material. In the early 2000s, titanium and zirconium were adopted as the coating materials. Electrochemical inertness of such metals as aluminum, titanium and zirconium makes a good diffusion barrier for the cathode.

In the late 1990s and early 2000s, competition for high energy capacity batteries was getting dangerously tight. Finally, drastic reduction in the design margin for the unreasonable increase in the energy capacity led to continuous field incidents. The effort to resolve this problem made battery industry turn its attention to the high capacity lithium nickel oxide, but battery engineers hesitated to employ lithium nickel oxide because it became extremely unstable when it was exposed to high temperature overcharging condition. When the excessive lithium atoms are removed by overcharging, nickel atoms in the nickel layer are apt to move to the lithium layer, in the course of which the layer structure is transformed into a more stable cubic structure and redundant oxygen atoms are spewed out to the electrolyte. In case the electrolyte reaches the flash point at the point of oxygen's spewing out, the electrolyte is likely to catch fire. For the purpose of preventing this phenomenon, other metal atoms are doped to hold nickel atoms in the metal layers firmly. Cobalt, manganese and aluminum were adopted for this purpose.

There are various combination of nickel based cathode materials, among which NCM is the most commonly used in the xEV application. The design concept in conjunction with NCM is to combine the advantages of LNO, LCO and LMO: energy capacity of LNO, power capability of LCO and the safety of LMO. It was expected that blending of these three materials would create the cathode material with combination of advantageous characteristics. However, unfortunately mechanical blending of three materials produces the material in which the disadvantage of each material is combined unfavorably. So, co-precipitation method was introduced to make the cathode material which exhibit combined advantages of three materials.

In the early 2000s, Sanyo, the top lithium ion battery maker at that time, made lithium ion batteries for power tools using lithium iron phosphate (LFP) cathode material. After the test of LFP batteries, engineers of a power tool maker were so impressed by Sanyo's LFP batteries that they announced that they wouldn't accept any other batteries than LFP batteries. This made lithium ion battery industry get interested in lithium iron phosphate cathode materials. Safety of lithium iron phosphate ($LiFePO_4$) cathode materials comes from the olivine structure and PO_4 tetrahedron. Oxygen, which is considered as a bad agent, is trapped doubly by the olivine structure and the tetrahedron alike, so excellent safety is maintained. At first, LFP could not be used as the cathode materials because of the poor electric conductivity, which was resolved by Hydro-Québec of Canada by applying carbon coating on the powder surface.

Low cost and good safety were considered as the major advantages of LFP cathode materials over other materials such as NCM and LMO. But low cost has not been realized yet. As regarding the safety, good safety becomes meaningful only when it is reflected in the cost of power packs. Only when LFP cathode material enables us to make cost-effective power packs thanks to the good safety, a high safety feature becomes meaningful business-wise. Therefore, the future of LFP batteries depend on whether or not the drastic reduction of the power pack cost due to the high safety features of lithium iron phosphate cathode materials is feasible.

The development trend of cathode materials is to increase the specific capacity by increasing the nickel content from 50% to 60-70% in NCM cathode materials. Another approach of NCM cathode is to increase the manganese content from 20% to 30-50% by the concomitant reduction of the cobalt content not only to improve the safety but to avoid the supply issue related to the cobalt metal. Cobalt mines are located mainly in Central Africa where the political instability has been a constant problem. Although the LFP battery exhibits excellent safety features, its low voltage of 3.2V puts a restriction on the application area. In order to resolve this issue, attempts have been made to increase the voltage by replacing iron atoms with manganese atoms.

Major suppliers of cathode materials are Umicore of Belgium, Nichia of Japan and L&F and Ecopro of Korea. The critical ingredient of success in the cathode materials makers is to find out how to collaborate with battery makers.

(2) Anode materials

Carbon is made by burning the precursor. Depending on the hardness of a precursor, it is called hard or soft carbon respectively. Graphite can be made from soft carbon by heat treatment called graphitization, but hard carbon cannot be converted to graphite. This is because the regularity of the crystal structure of soft carbon is more pronounced than that of hard carbon. Sony adopted hard carbon in the lithium ion battery in 1991 because hard carbon had higher specific capacity than soft carbon. The graphite made from soft carbon is an artificial graphite in contrast with a natural graphite mined from a graphite mine.

As regards the crystal structure of graphite, graphite has the layer structure like LCO. In each layer, six carbon atoms form the hexagon structure, which can hold one lithium atom, forming LiC_6. Graphite accepts electrons and lithium atoms from the outside in charging, forming the LiC_6 and releases electrons and lithium ions in discharging, which is called the intercalation. The basic concept of the intercalation is that graphite acts like a metal electrically. The higher the graphitization temperature is, the higher the energy capacity of graphite is by the increase of the crystallinity of graphite. Graphitization temperature of 3000°C is commonly practiced for this purpose. High capacity graphite is so soft that it tends to break in the roll pressing process. In order to strengthen its resistance to breakage, a blending method is implemented, in which high capacity graphite is blended with high strength graphite to acquire the strength to withstand the rolling pressure.

Natural graphite is mined in a graphite mine. Natural graphite is a cost-effective anode material, but introduction of natural graphite complicates the production methodology because of the gas evolution from the SBR binder. Precharging and high temperature aging process are adopted to remove the gas before sealing. In the xEV application, it was found that lithium is prone to be precipitated on the anode surface at low temperature charging. This phenomenon is caused by the abrupt increase in the interfacial resistance at low temperatures. The battery industry was not aware of this phenomenon because charging temperature of portable electronics was above 0°C. In effect, the xEV was the first application in which low temperature charging was required. Replacement of graphite by carbon or the addition of soft carbon to the graphite alleviated lithium precipitation, so the interest in the soft carbon anode was rekindled in the xEV application.

Silicon anode has been developed since the early 2000s because of its potentials as the high capacity anode materials. However, promising as it may look in terms of the energy capacity, enormous volume change of silicon during charging and discharging remains as the barrier to commercialization. Specific capacity of silicon is 4,010 mAh/g which is more than ten times higher than that of graphite, 372mAh/g. However, the volume change in charging and discharging is 300-400%. In graphite, the volume change is 12%. Development activity is naturally focused on finding the way to reduce the volume change of silicon. One way to reduce the volume change is to control the morphology of silicon. For example, the growth of silicon in the form of a tube can reduce the volume change. Doping of iron and titanium can be also helpful to control the volume change. Another way to alleviate the enormous volume change of silicon is to make the composite of silicon oxide to graphite. In spite of all the efforts to control the volume change of silicon, the current state is to add small amount of silicon (3-5%) to graphite for the slight improvement of energy capacity.

Fast charging and a long cycle life are important requirements of xEVs. Lithium titanate (LTO: $Li_4Ti_5O_{12}$) is the anode material that fits the bill. Volume change of LTO in charging and discharging is only 0.1-0.2%. Because of the extremely low volume change, it is called the zero-strain crystal structure. The feature of zero-strain crystal structure in combination with nano powders, whose specific surface area is $100m^2/g$, gives LTO batteries the fast charging capability. However, the LTO battery is susceptible to swelling. Swelling is caused by the reaction of LTO with the electrolyte, as a result of which such gases as hydrogen, carbon monoxide and carbon dioxide are generated. The crystal structure of LTO is spinel structure like LMO. The spinel crystal structure is vulnerable to swelling due to the position of the crystal plane. Altairnano and Toshiba Battery resolved the swelling problem by the surface treatment of the LTO powder. Either carbon coating or formation of solid electrolyte inter-phase (SEI) film on the LTO surface was proven to be effective for the prevention of the swelling. Another issue of the LTO battery is the low voltage. The maximum average voltage of LTO batteries is 2.4V.

With regard to the graphite supplier, vertical integration such as precursors and graphite mines is the important factor for success, so Japanese makers have been dominating the market. Major suppliers of graphite are Hitachi Chemical, Nippon Carbon and JFE Chemical of Japan. BTR Energy of China has joined the group of graphite suppliers.

2) Assembly materials

(1) The separator

The separator is classified as the dry and wet process separator depending on the manufacturing method. In a dry process separator, stretching force is applied in the rolling direction for the formation of air pores, as a result of which it is vulnerable to puncture for the low width direction strength. On the other hand, in a wet process separator biaxial stretching force is applied to give isotropic mechanical properties to the separator, so it has a good puncture resistance. As regards the pore shape, it has some bearing on the manufacturing process. The dry process separator has straight pores, which is in contrast with tortuous pores of a wet process separator. Lithium ions can move fast through straight pores, so the separator made by dry process exhibits a good power capability. Because of the simplicity of the manufacturing process, the dry process separator is cheaper than the separator made by wet process.

In 1990s Tonen and Asahi Kasai of Japan produced a wet process separator whereas Ube of Japan and Celgard of the USA produced a dry process separator. As the leader of a lithium ion battery business in the 1990s, Sanyo wanted to combine the advantage of both wet and dry process separator: high power capability due to the straight pores and high puncture strength by the increase of the width strength. Based on this design concept, Sanyo collaborated with Asahi Kasai to develop the hybrid separator using wet process.

GS-Melcotec of Japan, a thin prismatic lithium ion battery maker, employed a 18 micrometers separator made by dry process in the prismatic batteries to top its competitors in energy capacity as well as the cost. At that time, a 20 micrometers separator made by wet process was commonly used. GS-Melcotec wanted to kill two birds with one stone by adopting a low cost thin separator made by dry process. However, unfortunately GS-Melcotec's batteries had brought about field incidents because of the separator puncture. Taken aback by the unexpected field incidents, GS-Melcotec took a drastic measure. It changed the cell design rule not to use the separator made by dry process regardless of the thickness of the separator to prevent the outbreak of the field incidents. This decision was a fatal blow to separator makers using dry process. All of a sudden, Asahi Kasai and Tonen emerged as the major suppliers of the separator, and the separator market became a seller's market. Battery makers had a tough time to acquire sufficient quantity of the separator because of the limited production capacity of Tonen and Asahi Kasai.

In the midst of the unstable supply of the separator due to the field incidents of GS-Melcotec, Samsung SDI's battery engineers took their chances to employ the 18 micrometers separator made by dry process in their 18650 cylindrical batteries. These cells had raised serious field incidents all over the world for six months in 2003. It started in May, 2003 during the laptop computer aging process in the laptop computer plant located in China. Aging of laptop computers is the process in which we connect the power pack to the laptop computer body and conduct the cycling test twenty times to make sure that there won't be any events like fire or explosion during its usage in the field. In the aging process, one power pack was burnt after completion of 12 cycles, which was a new experience for the laptop computer makers at that time. After the burning event of power packs during the laptop computer aging, several field incidents broke out in Spain and Brazil and other parts of the world. Nonetheless, the battery maker as well as the laptop computers maker considered the field incidents as the isolated events in which hot weather might be the cause of fire or explosion. However, sporadic field events were developed into field incidents that were taking place uncontrollably all over the world regardless of the weather or climate.

The battery industry also took this incident seriously because of the involvement of the leading laptop computer maker, HP. In addition, frequency of incidents aggravated the seriousness of the field incidents. Laptop computers were burnt or exploded every week all over the world for six months. Flabbergasted by the frequency and geographical spectrum of the field incidents, chairman's office of Samsung group stepped in to investigate the cause of incidents and eventually resolve the issue. People in the battery industry were beginning to acknowledge that lithium ion batteries could actually burn or explode in a real environment.

It was hard to put our fingers on what caused the fire and explosion of laptop computers because these field incidents of 18650 cylindrical batteries resulted from the combination of several wrong things: wrong use of materials, improper cell design and mistakes in manufacturing. However, there is no doubt that negligence of management sparked the field incidents. This battery company made a mistake in expanding its production capacity without concomitant increase of management capability. As for the materials, the separator was the source of the problem. In order to make a three layer dry separator, we use polyethylene and polypropylene as raw materials. Celgard, the US separator maker, made a mistake in the quality control. Its quality control system failed to screen out the defective raw polymer. Introduction of polymer with molecular weight outside the specified range by mistake led to the non-uniform separator, the thinnest part of which was as low as 16 micrometers.

As regarding the cell design, the battery maker decreased the n/p ratio down to 1.03-1.06, which was low enough to deposit lithium metal on the surface of the anode unless electrode thickness is uniform. At that time, most of the Japanese battery makers stipulated the minimum n/p ratio to prevent the foolhardy try of a battery designer. The commonly accepted minimum n/p ratio at that time was 1.12. Considering the commonly accepted minimum n/p ratio, n/p ratio of 1.03-1.06 was a ridiculously low figure. Nevertheless, managers of Samsung SDI's battery business division didn't recognize the potential danger associated with the highly risky cell design. On the top of everything, the battery business manager pushed the cell designers to the limit to design batteries that would enable Samsung SDI to compete with leading Japanese battery companies in terms of energy density and cost. As a result, their cell design was in a touch-and-go state.

Battery manufacturing engineers should have controlled the electrode thickness as accurately as possible in the coating process to alleviate the danger accruing from the stringent n/p ratio, but they acted otherwise. They used a new coating machine installed in the new battery plant without sufficient verification. Battery business manager of that company urged engineers to reduce the verification period drastically. Pursuant to the pressing order of the battery business manager, engineers in charge of the verification of a newly constructed plant skipped most of the crucial steps in verification. They thought that scrupulous verification might not be necessary because the coater installed in the new battery plant was identical to the first one. Concerned about the potential risk associated with the highly risky design of 2.2Ah 18650 cells, the battery business manager strengthened the chemical analysis activity by increasing the frequency of the electrode analysis to make sure that uniform electrode is made in a new plant. Excessive orders from the battery business manager without the increase of workforce made the chemical analysis team cook up the data. They used the same analysis data day in day out by changing the time and date.

After the storm of field incidents blew over, there was an internal auditing to figure out what went wrong, but Samsung SDI and Samsung group failed to eradicate the root cause of the field incident. In effect, they couldn't even figure out what made the lithium ion battery burn or explode in the field. The field incident mentioned above left a permanent scar not only to the pertinent battery company but to the whole lithium ion battery industry. The argument that the lithium ion battery is safe because lithium exists in the form of ions instead of atoms was demolished by the extensive field incidents.

Surprised by the unexpected field incidents, HP, the leading laptop computer maker, made a rule not to use the lithium ion batteries whose separator thickness is below 20 micrometers. Turning to the field incident mechanism of this battery company, hard shorts may have occurred in the cells when the locality of protrusions on the electrodes due to the non-uniform thickness of the electrodes coincided with the thin part of a separator.

As regards Samsung SDI, thirteen years after the field incident in HP's laptop computers, field incident was repeated in Samsung Electronics' smart phone, Galaxy Note 7. Failure to eliminate the root cause of the field incident in 2003 made Samsung SDI repeat the same mistake in 2016.

With regard to the competition dynamics of separator makers, Tonen and Asahi's dominance in the separator market was finally weakened by SK's entry into the market with the wet process separator. In pouch batteries, rectangular shaped electrodes are made by stamping procedure. Edges of stamped electrodes could be the origin of ignition as the burr or sharpness of the edge is prone to puncture the separator. To prevent the separator damage at the edge of the pouch battery electrodes, ceramic coating on the separator is implemented for the sake of robust design. SK innovation and LG Chemical are famous for the ceramic coating technology.

In the investigation of field incidents, the separator is frequently blamed for the fire or explosion, so the separator makers should have the capability to deal with the field incident financially. Small companies, no matter how good their technology is, are not suitable for the separator supplier. Major suppliers of a separator are Asahi Kasai, Toray Tonen of Japan and SK innovation of Korea.

(2) The electrolyte

Determination of the electrolyte is the last step in the battery development procedure, so knowledge on the electrolyte is instrumental in the development of batteries. According to the investigations on the field incident, many field events resulted from the lack of knowledge on any specific materials such as a separator, carbon black and the like. Therefore, comprehensive knowledge of the battery materials is required not only to develop batteries with the balanced cell performance but to guard against the field incidents.

Current trend of the electrolyte development is to use additives to cope with the specific requirements, but indiscreet use of additives without sufficient knowledge on the side effects can be hazardous. In general, additives influence the SEI film, so they usually decrease the cycle life compared to the batteries with the electrolyte devoid of additives. Some additives discolor the electrolyte and electrodes severely, the effect of which has not been completely understood yet. Therefore, additives should be used with careful consideration on the side effects.

The electrolyte is the first battery material that was commercialized in Korea. Cheil Industries, subsidiary of Samsung group and currently identified as Samsung SDI after being merged into Samsung SDI, commercialized the electrolyte with the help of Samsung SDI. At that time, Samsung SDI had a close relationship with Mitsubishi Chemical of Japan for the purchase of the electrolyte. When Cheil Industries commercialized the electrolyte and supplied the electrolyte to Samsung SDI, Mitsubishi Chemical felt betrayed. Mitsubish Chemical thought that Samsung SDI acquired the technology associated with the electrolyte through numerous contacts with Mitsubishi Chemical and handed over to Cheil.

Based on the suspicion and reasoning, Mitsubish Chemical made a strong protest to Samsung SDI against the wrongful use of technical information acquired in the business meetings. However, Samsung SDI maintained its policy not to make any response to the protest. Aggravated by the no-response policy of Samsung SDI, Mitsubishi Chemical was determined to get even. Mitsubishi Chemical approached LG Chemical, a rival company of Samsung SDI, and offered to show its battery business activities to the key members of LG Chemical. Although LG Chemical was bewildered by Mitsubishi Chemical's offer, it had no reason to turn down the offer. Three people from LG Chemical were invited to Mitsubishi Chemical. Mitsubish Chemical showed the utmost courtesy to LG Chemical by picking up the tab arising from the trip to Japan among other things.

Several years after the incident, Samsung SDI asked Mitsubishi Chemical to supply NCA and NCM cathode materials for the development of xEV batteries. Mitsubishi Chemical gave the cold shoulder to Samsung SDI. Samsung SDI was beginning to acknowledge that it should mend the fences with Japanese battery materials makers including Mitsubishi Chemical. One of the weakness of Samsung SDI's battery business in the 2000s was that its dependence on the outside source in the battery materials was too high, so Samsung SDI couldn't afford to be isolated in the battery materials industry. Fortunately there were some battery engineers from Japan at Samsung SDI. Japanese battery engineers who were working with Samsung SDI came forward to improve relations with Japanese battery materials companies including Mitsubishi Chemical by taking advantage of their connection and network. Thanks to the effort of Japanese battery engineers of Samsung SDI, bad blood between Samsung SDI and Mitsubishi Chemical finally came to an end and, more importantly, the relationship with major Japanese battery materials suppliers was also improved significantly. At any rate, both Samsung SDI and Mitsubishi Chemical need each other to grow together in the battery industry.

Major supplier of the electrolyte are Mitsubishi Chemical of Japan, Guotai Huarang of China and Panax-Etec of Korea.

13
:
Future battery business

1) Trend of xEV batteries

One of the most effective ways to expand the xEV market is that a battery maker collaborates with an automobile maker to sell its batteries to other automobile companies after the development of the battery that meets the wish list of the automobile industry. This marketing methodology is called the co-marketing. Co-marketing is the starting point of the battery standardization. Once the battery is standardized, automobile makers will be able to reduce the cost dramatically by the improvement in the efficiency in the distribution channel.

Similar methodology had also been used in the battery materials in the 1990s in Japan. When a battery maker developed new battery materials in collaboration with a materials company, the battery maker allowed a materials company to supply new materials to other battery makers, and it had the benefit of purchasing new materials at a low cost. In so doing, lithium ion battery industry was able to grow the market in a short period of time. In actuality, the way Japanese lithium ion battery makers handled new technology played an important role in growing lithium ion battery business. They set great store by growing together in a highly competitive battery market.

In contrast, western companies which developed lithium iron

phosphate (LFP) cathode material demanded payment of a ridiculously high patent fee, as a result of which the LFP battery market had shrunk. In effect, request of payment of a high patent fee by companies which did not participate in the lithium ion battery business was the major reason why battery companies were so reluctant to adopt lithium iron phosphate cathode materials in their batteries. China and Korea were excluded in the countries in which LFP patents were registered, so battery companies in China and Korea can produce LFP batteries without the danger of the patent infringements as long as the vehicles installed with LFP batteries are used inside China or Korea. For the establishment of the reasonable patent fee, it is necessary that companies which own the patent rights are directly involved in the battery business. Otherwise, they will use the patent as the entry barrier by demanding the payment of a patent fee high enough to make the battery business economically unfeasible.

Asahi Kasai of Japan, which had the original patents of lithium ion batteries and formed a joint battery company with Toshiba Battery, requested the payment of its patents only after Asahi Kasai terminated the lithium ion battery business. Japanese lithium ion battery companies including Asahi Kasai may have thought that unreasonable request of payment of the patent fees without regard to the cost structure of the battery would prevent the expansion of the lithium ion battery market. At any rate, the essence of the competition of lithium ion batteries is to take one step ahead of others, not to drive away competitors from the market.

As regards the form factor of xEV batteries, some automobile companies, especially German automobile makers like BMW, have been particular about the robustness of lithium ion batteries because unlike small lithium ion batteries used in mobile phones and laptop computers, batteries should last for over ten years under harsh conditions. With regard to the robustness of the battery, prismatic batteries using the metal can and laser welding sealing exhibit better performance than the pouch batteries. However, pouch batteries have many advantages over prismatic batteries in other areas such as heat dissipation and power capability. In addition, unlike small prismatic lithium ion batteries whose cell design had already been perfected, the large prismatic batteries for xEVs need to be improved and refined further in terms of the cell design to insure reliability.

The ease of the acceptance of LG Chemical's pouch batteries in the electric vehicle market is due to the robustness of LG Chemical's pouch batteries. When LG Chemical's battery engineers designed the pouch batteries in 1998-1999, they focused on improving the robustness of the conventional pouch batteries, which explains why the LG Chemical's pouch battery is hard to the touch. The principle behind LG Chemical's pouch battery design was to make a hybrid battery of the pouch and the metal can prismatic battery in terms of the robustness and ease of handling in the pack assembly process. In small lithium ion batteries, cylindrical and prismatic batteries are more cost-effective than pouch batteries in terms of the production speed and automation. However, as the size of the battery becomes large, the gap between these batteries becomes small simply because of the enlargement of the size. By and large, competition between prismatic and pouch batteries is far from over in the xEV battery market.

Although LG Chemical is a chemical company in nature, LG Chemical has more experience in the pack and BMS design than Samsung SDI, an electronic company within Samsung group. LG Chemical's association with CPI and Aerovironment of the USA reflects its effort to reinforce the capability in the pack and BMS design. When LG Chemical developed the xEV batteries in 2000-2003 in collaboration with automobile companies, automobile companies didn't have the cell testing equipment. They only had the test equipment of the 360V pack, so early movers like LG Chemical had to supply battery samples in the form of 360V power packs. Therefore, early movers were forced to develop pack and BMS by necessity. LG Chemical made intensive investment in the pack and BMS on the assumption that they would be able to supply the xEV batteries in the form of 360V power packs integrated with BMS. However, contrary to the LG Chemical's expectation, Hyundai Motor wanted LG Chemical to transfer the pack and BMS technology to Hyundai Mobis, the automotive parts company within Hyundai Motor group, and supply xEV batteries in the form of individual cells to Hyundai Mobis for the assembly of the xEV power pack. Needless to say, there was an intensive and lengthy debate between two parties regarding who does what. As a compromise to the conflict in interest, they agreed to form a 51:49 joint company of Hyundai Mobis and LG Chemical in November, 2009. This is how HL Green Power, the xEV power pack assembly company, was created in 2010. HL Green Power acted as a kind of buffer zone between Hyundai Mobis and LG Chemical and, interestingly enough, battery engineers who used to work with Samsung SDI, a rival company of LG Chemical, had played an important role in the operation of HL Green Power.

LG Chemical was keen to find a market in which it could utilize the pack and BMS technology. It found a market which is in a nascent stage. It was the ESS market. LG Chemical developed ESS packs using pack and BMS technology accumulated through the development of xEV batteries. This is how LG Chemical became the front runner along with Samsung SDI in the ESS market. In effect, it was LG Chemical and Samsung SDI that created and grew ESS market in 2009-2011. It is interesting to know that the conflict between the battery and automobile companies concerning who would take ownership of power packs played an important role in creating and growing the ESS market.

Samsung SDI was not an early mover in the xEV battery industry. When Samsung SDI started the development of xEV batteries in 2004, most of automobile companies had the capability to develop pack and BMS and, more importantly, they had the cell evaluation capability. Samsung SDI wisely decided to focus on the cell development without any development activity of a pack and BMS to shorten the development time frame. This gave Samsung SDI the chance to catch up early movers like LG Chemical. As regards SK, SK devoted more efforts to the development of pack and BMS for HEV to get over the intrinsic limitation of the chemical company.

After termination of the PDP business, Samsung SDI moved most of development forces of PDP to the battery business to develop the pack and BMS in 2009-2012. Therefore, the technology base associated with pack and BMS of Samsung SDI is rather obscure as compared with LG Chemical and SK innovation. After establishment of SBL, a joint company of Samsung SDI and Bosch, Samsung SDI and SBL agreed on the market they could penetrate into. Pursuant to this agreement, all the moving applications belonged to SBL and Samsung SDI was allowed to develop only the stationary applications. Hence, Samsung SDI had to look for stationary applications to maintain the continuity in the development of large lithium ion batteries.

The ESS was the stationary application market which Samsung SDI found promising. In general, we use same cells in both the xEV and the ESS applications. Therefore, development of ESS batteries is centered on the development of the pack and BMS. Samsung SDI's focus on the ESS battery business due to the restriction on the development of the vehicle application put Samsung SDI in the forefront of the ESS business in 2010. KoKam was in the same situation as Samsung SDI after it transferred the xEV battery technology to Dow-KoKam. Naturally KoKam participated in the ESS business. LG Chemical, Samsung SDI and KoKam along with BYD of China formed the top four group in the ESS business. However, it should be noted that the ESS is the market in which reused xEV batteries should be used for cost competitiveness.

In the automobile industry, some people think that the pack business should belong to the automobile companies because vehicle information, some of which are considered confidential, is required to design a proper BMS. However, in accordance with the value chain of a xEV battery business, the xEV pack business will be likely to be shifted to the battery makers like what had happened in mobile phones and laptop computers in the 1990s.

As regarding the partnership with automobile companies, no battery companies in Korea have established the firm ground in terms of the co-marketing yet. Therefore, future competition will hinge strongly on which battery company will develop a working relationship with major automobile makers like Sanyo-Nokia relationship in the area of mobile phones.

With regard to the battery business direction, Korean battery companies seem to play with two options: to remain as the battery makers like Japanese battery companies or to jump to the electric vehicle makers like BYD of China and Tesla Motors of the USA. If they choose the former, they may have to form a joint company with major automobile makers for the expeditious pursuit of the business. In contrast, if they prefer to adopt the latter, chances are they will form a joint company with automotive parts firms for the implementation of vertical integration. Investigation of the path which major Korean battery companies had taken since the early 2000s indicates that major Korean battery companies seem to prefer to take the latter, the example of which can be found in the formation of joint companies with automotive parts firms of Germany. However, it's not clear whether forming the joint companies reflected their effort to implement the vertical integration for the electric vehicle business or it was just a strategic blunder.

The xEV is still in the development stage in terms of the cost structure. Automobile companies are free to stick to the conventional automobiles unless electric vehicles do not give them a satisfactory profit. That's the difference between the small lithium ion battery and the xEV battery business. In the small lithium ion batteries used in mobile phones and laptop computers, the battery was the critical component without which portable electronics makers should discontinue their business. Since their dependence on the performance and safety of the advanced batteries was very high, electronic companies were so active and aggressive in growing lithium ion battery business. Unfortunately, that's not the case in the xEV business. In the situation like this, a newcomer in the xEV industry like Tesla Motors may play a pivotal role in expanding the market. In this regard, the strategic movement of Korean major battery makers such as Samsung SDI, LG Chemical and SK innovation is critical to the future xEV market because they have the capability to become xEV vehicle makers and eventually change the competition dynamics of automobile industry.

2) Cost target of xEV batteries

One of the most important issues in the battery business has been the cost target. The current battery cost of electric vehicles is about \$300/kWh on a pack basis, which is equivalent to \$230/kWh on a cell basis. When the goal of \$200/kWh (cell: \$155/kWh) is reached, major battery makers will be likely to make profits from the battery business. If the cost reaches \$100/kWh (cell: \$80/kWh), the electric vehicle can compete with the conventional IC (internal combustion) engine vehicle without help of government subsidy. The main flow for the cost reduction activity encompasses improvement in the cell design and increase in the specific capacity of materials. In addition, reduction of the number of cells used in a pack will be helpful to reduce the pack cost because the complexity and the cost of a pack increases with the number of cells. If the goal of \$100/kWh is achieved, chances are most battery makers which survived the tight competition will make profits from the battery business.

As regards the number of cells in a pack, 96 prismatic cells are connected in series to form a 360V power pack, and 288 cells are connected in pouch batteries, in which 3 cells are connected in parallel and 96 cells are connected in series. The pouch case is fabricated by pressing the aluminum sealing paper on the die, so pouch batteries have the limitation to make a large and thick cell. The maximum thickness of the pouch case we can make without tearing the edges of the pouch case is 6mm. When we combine two pouch cases together, we can make the pouch case with the thickness of 12mm, which is the maximum thickness of pouch batteries. Because of the restriction on the thickness, parallel connection is required to meet the required capacity.

In 18650 cylindrical batteries used in the electric vehicle of Tesla Motors, 7,104 cells are connected to form a 360V pack. Tesla Motors adopted liquid cooling of cells in a 22V module to insure uniform cell temperatures. Temperature and temperature distribution are the most important parameter we have to control in the design of a power pack. In prismatic and pouch batteries, when temperature inside a pack reaches 50°C, a cooling fan is turned on to cool down the temperature of cells. Cooling air passes through the main channel and is distributed to individual cooling channels to insure the uniform temperature distribution.

According to the roadmap of LG Chemical and Samsung SDI, the cost target for 2020 is $100/kWh on a cell basis. LG Chemcal is to supply cells to GM at the price of $145/kWh on a cell basis and Samsung SDI increased the energy density by increasing the cell capacity from 60Ah to 94Ah for BMW i3. Samsung SDI changed the cathode materials from LMO+NCM to NCM+NCA. Samsung SDI and LG Chemical argue that they can supply their cells at $100/kWh on a cell basis and make profits by 2020. However, their cost structure may not allow them to achieve their ambitious goal. For one thing, the operating rate of the factory for LG Chemical's battery business was 62% in 2015 and just above 65% in 2016 because of the overinvestment in 2009-2012. In addition, the materials cost accounts for over 70% of the total cost. Considering the materials cost is below 45% in small lithium ion batteries, more time is needed to stabilize the cost structure. Therefore, more realistic goal for 2020 is likely to be in the neighborhood of $200/kWh on a pack basis.

There are a couple of ways to reach the $200/kWh goal from $300/kWh. First, we can increase the energy density of the battery by replacing NCM+LMO cathode with NCA+NCM (50:50) like Samsung SDI did to satisfy BMW's new requirements. Another method is to increase the specific capacity of NCM by increasing the nickel content in the NCM cathode materials like LG Chemical. Second, we can improve the cell design to contain more anode and cathode materials. If there were more room for the densification of the electrodes, the increase of electrodes density would be helpful to increase the energy density. Decreasing the volume of the inert materials like metal foil and the separator would give us more space for active materials, which increases the energy density of batteries.

With regard to the 18650 cylindrical cell, it has been on the fast track in ramping up the energy density because of the use of NCA. The pack cost of Tesla Motors is known to have reached $200/kWh, but the problem confronting the power pack of Tesla Motors is that the cost is beginning to level off. That's why Tesla Motors is trying to switch to a bigger cylindrical cell like 20700 cylindrical cell to reduce the cost of pack and BMS. The cost of pack and BMS increases with the increase of the number of cells. In addition, we cannot rule out the possibility that Tesla Motors change the form factor from small cylindrical cells to prismatic batteries after reaching the cost target of $200/kWh.

Economies of scale and the new anode materials will be the main driving force for the achievement of the final goal. Component suppliers of lithium ion batteries are divided into two groups depending on the size of the component supplier. The anode materials and a separator are supplied by big companies. As for the anode materials, vertical integration is required for the efficient acquisition of precursors. Hitachi Chemical, Nippon Carbon and JFE Chemical of Japan are in possession of the precursors internally. Whenever there are field incidents like fire or explosion of batteries, the separator is the first battery component that is likely to be blamed for the incident. Therefore, the separator makers should have the capability to deal with field incidents systematically and financially. That's why big companies such as Asahi Kasai, Toray Tonen of Japan and SK innovation of Korea participate in the separator business.

Unlike the anode materials and a separator, the suppliers of cathode materials and the electrolyte are rather small. Suppliers of the cathode and the electrolyte need to collaborate with battery makers to meet the needs of battery makers, so close connection with battery makers is essential to maintain the market. For the collaboration with battery makers, small companies are more efficient than large companies, so small companies with the exception of Mitsubishi Chemical have conducted the cathode and the electrolyte business. However, in the automobile battery, it is expected that there may occur some changes in the cathode materials and the electrolyte makers by the process of merger as the market expands. In this process, the cost can be reduced by economies of scale.

The biggest obstacle to the cost reduction is the fact that energy density of lithium ion batteries incorporated with graphite anode and NCM cathode seems to be approaching the practical upper limit although we don't know what the upper limit of energy density is until field incidents actually happen. Therefore, development of new materials is necessitated to get over the cost barrier. In this regard, it is essential that we develop the technology by which new materials like the silicon anode can be applied to lithium ion batteries. It should be noted that the cost target of $100/kWh is based on the assumption that there would occur the technological breakthrough in the battery materials within 5-7 years. Considering the time frame required to develop new materials, the expected time of the final cost target of $100/kWh will be in the neighborhood of 2025. The time has come that we may have to pay more attention to the development progress of the silicon anode with enormously high specific capacity.

3) A new method

The battery business without a battery plant is the battery networking business. As the lithium ion battery manufacturing technology becomes more and more commonplace, the software which consists of battery design, formula and manufacturing recipe along with the market information will be more important than the hardware which is merely the production activities. In the end, the time will come that we may not need the manufacturing plant to enter into the battery business. If this is the case, the power pack business could possibly become the center of the value chain of the battery business instead of the cell manufacturing business simply because of closeness to the market. In case the power pack makers have the capability to design the cell with the sales point that differentiates themselves from other battery makers, they may be able to compete with the traditional battery makers without bothering to have their own manufacturing plant. By and large, the focus is being shifted from hardware to software in the lithium ion battery industry and this trend will become more and more obvious with the maturity of the battery manufacturing technology.

Value chain of the battery business flows from battery materials to cell manufacturing to the pack assembly business. Among these, the cell business has been the center of the value chain. However, things have been changing steadily since 2010. Problems associated with oversupply and the malignant stock of xEV cells caused by the excessive investment in the xEV battery plant accelerate this trend. Even the small pack companies can be a battery maker if they acquire the capability to design the battery, which is beginning to take place in the golf cart market.

The networking business may be able to change the major players in the xEV batteries. Automotive parts companies like Bosch, Continental, Johnson Control and Hyundai Mobis will be likely to emerge as formidable players in the battery market. Their acquisition of battery technology from major battery makers made it possible to compete with traditional battery companies in the same market. If they use networking business tools to compensate for the lack of experience in the manufacturing of lithium ion cells, chances are they can beat the conventional battery makers in the same market. In the situation like this, the competition hinges on who has the better design skills and creative ideas rather than who makes better cells.

Introduction of the concept of the networking business makes it possible for the companies with creativity and ingenious ideas such as Google and Apple to enter into the battery business without repeating the same mistakes as Eveready and A-123 had made. Companies like Google and Apple can be very successful and eventually may be able to lead the world xEV battery and vehicle market if they are patient enough to pursue the battery business step by step. Based on the history of success and failure in the battery business, we can possibly come up with an idea as to how we can enter into the battery business safely and efficiently. Here is one of the success scenarios. In order not to follow in the footsteps of Eveready of the 1990s and A-123 of the 2000s, both of which had focused their attention exclusively on the manufacturing of cells and had tried to change the competitive competition too rashly, it is important to take a stepwise approach.

First of all, we need people with experience in the major battery companies for the design of the target battery. When designing the battery and making the production recipe of the battery are completed, we should find the battery makers that can make cells based on our recipe, and then we set up a power pack assembly plant. In-house power pack assembly line is necessitated since power pack business interacts with the automobile companies directly, so it acts as the information center of the whole business.

After becoming a battery maker using the battery networking business tool, the next step is to be a xEV vehicle maker and additionally jump to a leading battery maker by introducing new battery technology, the method of which is the simple extension of the networking business described above. Simple as it may sound, only a few companies have the capability to go ahead with the battery business utilizing the networking business tool. Hanwha Chemical of Korea planned a battery networking business in 2013, but the activity got stopped by the abrupt decision to stop all the activities associated with batteries in January, 2014. The networking business is just one example of the future battery business. By the introduction of new methodology like the networking business, the battery industry will be diversified significantly.

In addition to the battery networking business, collaboration of Tesla Motors with Panasonic of Japan will have influence on the competition landscape. Gigafactory to be constructed in Nevada will be the biggest battery plant in the world. Collaboration of Tesla Motors with Panasonics reminds us of the combination of Toyota with Panasonic in the 1990s and the collaboration of Nokia with Sanyo in the mobile phone market.

It is interesting to know how the lithium ion battery expanded the market horizon for the last twenty-five years. Lithium ion batteries beat nickel metal hydride batteries in portable electronics market such as mobile phones, laptop computers and the camcorder in 1990s. It was a surprise victory since many people in the battery industry thought that the nickel metal hydride battery, which was a year older than the lithium ion battery, would be more competitive in terms of the voltage compatibility, cell performance and cost.

In 2003, high power lithium ion batteries penetrated into the power tools market in competition with nickel cadmium batteries. The power tools market had been considered as the impenetrable market for lithium ion batteries because of the inability of lithium ion batteries to operate the electric motor until lithium ion batteries had the capability to discharge at high rate in 1999.

After acquisition of high rate discharge capability, the lithium ion battery expanded its market from power tools to the xEV market. From the mid 2000s, large format high power lithium ion batteries were beginning to capture the xEV market competing with high power nickel metal hydride batteries. In the 2010s, lithium ion batteries expanded into the field of the ESS which was created intentionally by Korean battery makers. In this way, the lithium ion battery has been expanding its marketing horizon since 1991.

You may be wondering what the next target market will be. The next target market for which the lithium ion battery is striving is the lead acid battery market, which requires large energy capacity as well as the cost competitiveness. Therefore, future development activities should include the development of large and cost-effective batteries. The maximum energy capacity of a single cell that lithium ion batteries can achieve is below 100Ah with the exception of large format plastic can batteries incorporated with lithium iron phosphate cathode materials. The maximum energy capacity limit of lithium ion batteries should be raised well above 100Ah to be competitive in the lead acid battery market.

When will we be able to use lithium ion batteries in the submarine? This is the question we should ask ourselves to figure out when and how lithium ion batteries will be able to expand the application area. If we can apply lithium ion batteries in the submarine, we can safely say that lithium ion batteries finally became the universal battery. The need to replace the conventional lead acid batteries with the advanced lithium ion batteries in the submarine is high for several reasons, one of which is that lithium ion batteries can extend the battery replacement time significantly. The lead acid batteries in the submarine should be replaced every four years. This replacement period can be extended to over ten years in lithium ion batteries. However, considering the battery requirement of the submarine battery, it would be a real challenging task to use lithium ion batteries in the close space of a submarine. Electrical requirement of the submarine battery is 10,000Ah and 900V. A single lead acid cell used in the submarine has the dimension of 30cm×30cm×150cm. The safety issue of lithium ion batteries is another concern. In short, we have a long way to go in terms of the battery size and the safety for the successful application of lithium ion batteries in the submarine. That's why the submarine battery is the ultimate goal lithium ion battery industry should be aiming at in terms of the cell size.

The most prominent advantage of lead acid batteries is the cost. Hence, the cost of lithium ion batteries should be slashed dramatically to compete with conventional lead acid batteries. As a way to reduce the cost, BMS and other pack parts should be simplified drastically although it may not be possible to completely eliminate them. In the late 1990s and early 2000s, the battery industry was keen to remove the PCM by using spinel manganese oxide but this effort was not successful at that time. The target battery was called a PCMless battery. Expectation of the possible simplification of BMS by the introduction of high safety lithium iron phosphate cathode materials was rather high, but contrary to our expectation, LFP batteries failed to reduce the pack cost.

Competition in the battery business will be very tight because of the oversupply and the possible full-scale involvement of automotive parts companies in the xEV battery market. In order to surmount the barriers ahead of us, it is advisable that battery companies concentrate their efforts on the expansion of the market by replacing the conventional lead acid batteries as well as the creation of the next-generation lithium ion batteries incorporated with new battery materials. Otherwise, battery companies might be overshadowed by their larger, faster competitors from outside.

Closing remarks

Sony created the lithium ion battery by making it operable in a real environment by the adoption of protection circuit module (PCM) among other things, and Sanyo improved its competitiveness significantly by increasing the energy density. Panasonic along with Sony and Sanyo expanded the marketing horizon from 3C application to power tools to xEVs to put lithium ion batteries one step closer to the universal battery. Korean companies such as Samsung SDI and LG Chemical played an important role not only in expanding the markets but in hindering Japanese makers from dominating the battery industry. In addition, BYD of China showed us how to cut some corners to make cost-effective lithium ion batteries.

Lithium ion battery technology which was once regarded as the restricted technology belonging to a few became so common that competition with battery makers from other industries would be inevitable. As an example, the likelihood of participation of automotive parts companies in the xEV battery market is getting bigger, which is the bottom-line consequence of declining competitiveness of lithium ion battery makers. Therefore, it is imperative that battery companies focus on the development of next generation lithium ion batteries incorporated with new materials to cope with the impending difficulty in the market.

Performance of Korean battery makers was more than impressive. In effect, they exceeded their expectations in terms of the market share.

However, the intrinsic problem associated with Korean battery industry is that all the companies produce only lithium ion batteries regardless of their business base. In order to provide Korean battery industry with the stability and balance for the sustainable competitiveness, versatility in the product portfolio is urgently needed. Hence, major Korean battery companies like LG Chemical, Samsung SDI and SK innovation should expand their business area in parallel with the systematic pursuit of xEV batteries.

Battery-related market will be expected to change its form to accommodate the future needs. There are a couple of factors that could change the competitive composition in association with xEV batteries. One of them is the movement of the major Korean battery makers. They can enter into the self-driving car business through the implementation of vertical integration. The other critical factor is the strategic decision of the US firms such as Google and Apple. They are likely to enter into the xEV battery business and subsequently a self-driving car business by taking advantage of their knowledge and experience of IT business by way of the battery networking business. Needless to say, movement of the companies like Google and Apple attracts the attention of the battery industry.

Japanese and Korean battery makers experienced field incidents in 2002-2006. The damage of the field incidents was so devastating that the change of the competition landscape was unavoidable. Explosion of Samsung Electronics's Galaxy Note 7 in August, 2016 flabbergasted the battery industry because there had not been any conspicuous field incidents for ten years. We may enter the era of the field incidents, so we should keep in mind that lithium ion batteries explode not only in the abused condition but also in the normal operating condition if the energy density is higher than the maximum upper limit.

The lithium ion battery is a potentially dangerous device, but the battery makers' compliance with the cell design rule guarantees the absolute safety under any circumstances.

Index

I

J

K

L

M

N

O

P

R

S

T

U

V

W

X

Z

Author's profile

Jun SONU

President of TOP21, Battery consultant

Academic background

- B.S. in Seoul National University, Seoul in Korea
  - Metallurgical Engineering (1980)
- M.S. in Wayne State University, Detroit in Michigan, USA
  - Metallurgical Engineering (1984)
- Ph.D. in University of Cincinnati, Cincinnati in Ohio, USA
  - Metallurgical Engineering (1988)

Work experience

- 2016-: Battery consultant, TOP21
- 2010-2013: Vice President in Hanwha Chemical
  - Development of xEV batteries
- 2002-2006: Vice President in Samsung SDI
  - Development of xEV batteries with Ford Motors
- 1996-2001: Senior Manager in LG Chemical
  - Business planning and development management of lithium ion batteries
- 1989-1995: Manager in LG Metals
  - Development of lithium polymer batteries with AEA of the UK